ELECTRICAL BREAKDOWN IN GASES

ELECTRICAL BREAKDOWN IN GASES

Edited by

J. A. REES

Department of Electrical Engineering and Electronics
University of Liverpool

A HALSTED PRESS BOOK

JOHN WILEY & SONS

New York – Toronto

1973

PHYSICS

© J. A. Rees 1973

First published 1973
THE MACMILLAN PRESS LIMITED
London and Basingstoke
Associated companies in New York Melbourne
Dublin Johannesburg and Madras

Published in the USA by Halsted Press,
a Division of John Wiley & Sons, Inc., New York

Library of Congress Cataloging in Publication Data

Rees, J. A comp.
  Electrical breakdown in gases.

"A Halsted Press book."
  1. Electric discharges through gases.   2. Ionization of gases.   I. Title.
QC711.R35   537.5'32   72-10338
ISBN 0-470-71335-6

Printed in Great Britain

Acknowledgement

The editor and the publisher wish to thank the publishers and authors who have kindly given permission for the use of copyright material included in this book.

Contents

Contents

Contents

Introduction

The electrical breakdown of gases has been studied in the laboratory since before the start of this century and it is sixty years since Sir J. S. Townsend wrote his monograph *The Theory of Ionization of Gases by Collision.* In that time the subject has been widely studied and a great deal of progress has been made in understanding many aspects of the problem. To select approximately twenty papers from the vast number that have been published in the past seventy-five years is a rather forbidding task, not simply because of the large number of papers available but also because the advances in knowledge that have been made have often come as the result of a number of good papers from various authors rather than from a single brilliant piece of work. Such a form of progress is very understandable in a subject as complex and diverse as the electrical breakdown of gases and it is often difficult therefore to determine which particular investigator first suggested the correct explanation of a particular phenomenon. The present selection is therefore a very personal selection and it is certain that others would have selected an entirely different set. The selection is intended primarily for final-year undergraduate students or first-year postgraduate students (and their teachers) as an introduction to the subject. Other research workers may be interested in re-reading the papers for pleasure.

Since the experimental facts and theoretical explanations on a particular topic are often better documented and more clearly expressed in an author's second or third paper on the subject I have not tried to always use the first paper every time. For example, paper 3 was not the first by Professor Raether on cloud chamber studies of avalanches but it is a good early paper nevertheless and will I trust give the reader an appreciation of the work described. In two cases I have gone further and have selected a review paper on a particular topic. For example, Professor Haydon's paper *Spark Channels* is a survey of a great deal of work by many authors and its inclusion might perhaps be criticised by some on these grounds. I make no apology for selecting it, however, since it is an excellent paper of its type and gives a more balanced view of the subject than could have been obtained by selecting one of the original papers cited in it.

The papers included in the first half of the book deal with electrical breakdown in uniform electric fields and in particular with the pre-breakdown processes leading up to breakdown. This is, of course, the simplest situation to study but is nevertheless a difficult problem that is still producing material for discussion. The classic experiment, still in use, is the so-called 'Townsend growth of current' experiment, described here in detail in paper 6. The experiment revealed the now famous equation

Introduction

$$I = I_0 \; \frac{\exp \alpha d}{1 - (\omega/\alpha)\{\exp(\alpha d) - 1\}}$$

for the spatial growth of current in a uniform discharge gap and confirmed Townsend's view that the basic parameter in these circumstances is the ratio E/p of the electric field strength to the gas pressure. It is difficult to realise now how significant the recognition of the latter fact was in the early work.

The Townsend theory of the breakdown process has not always been fully accepted for all conditions, e.g. appreciable overvoltages, and indeed is still questioned by many authors today. Some of the early objections are outlined in papers 2 to 5 and led to the alternative theories described by Professor Raether in paper 4 and by Professor Meek in paper 5. Paper 4 is not the first describing Raether's 'kanal theory' but does summarise clearly his early ideas. The theory described in paper 5 was arrived at independently of that described by Raether and grew out of qualitative proposals made by Professor Loeb and his co-workers as a result of their studies of breakdown in non-uniform electric fields. The two theories are very similar and are now referred to, loosely, as the 'streamer theory of break-down'. In the last twenty years the question 'Townsend theory or streamer theory?' has inspired a considerable amount of both experimental and theoretical work and has produced a great deal of new information about the basic ionisation processes involved in the breakdown mechanism. Papers 6 and 7 are examples of the efforts made in the early 1950's to determine whether there is a stage in the development of a pre-breakdown current where the steamer process takes over from the Townsend amplification process. There has not been space here to give one or more of the recent papers on studies of single and multiple avalanches in gases written by Raether and his colleagues in which the earlier cloud chamber studies described in paper 3 have been greatly extended by elegant oscillographic and image converter techniques. However, these methods and the results obtained have been summarised in Raether's book *Electron Avalanches and Breakdown in Gases* and should certainly be carefully studied. The papers (referred to elsewhere), by Professor Llewellyn Jones and his colleagues and more recently those by Professor Bruce's group and that of Professor Dutton and Professor C. Grey Morgan, should also be consulted as contributing greatly to the continuing effort. The papers on this topic have been studies of either the spatial growth to breakdown of small ionisation currents or of the temporal growth of a pre-breakdown discharge and in connection with the studies of temporal growth the papers by the late Professor P. M. Davidson (papers 8, 9 and 10) should be carefully examined. The papers are concise and make a contribution far greater than that suggested by their length. As indicated on page 83 the interested experimentalist will probably benefit from reading the explanatory sections of Grey Morgan's book *Fundamentals of Electrical Discharges in Gases* in conjunction with Davidson's papers.

Studies of breakdown in uniform electric fields would not be complete without studies of electro-negative gases, such as oxygen, and gas mixtures such as air, and papers 11, 12 and 13 illustrate the major influence electron attachment can have on the breakdown processes. Rather surprisingly it was 1953 before attachment was seriously allowed for in growth of current experiments. In more recent years Professor Frommhold, Dr. Schlumbohm, Professor Craggs and his colleagues, and others, have found it also necessary to allow for the reverse process of electron detachment in the analysis of their data. A considerable amount of important work on the breakdown of strongly attaching gases has also been carried out in the last fifteen years (see paper 13).

In the second half of the book, papers dealing with electrical breakdown in rather more specialised conditions have been considered together with papers on the development of the later stages of the breakdown process in which the current densities may be becoming high. The three papers 14, 15 and 16 describe the experimental features of spark channels as they develop from the time of initial breakdown, through a low current channel, to one of higher current and attempt to describe the build-up theoretically. The experimental techniques involved are fairly sophisticated and more information can be expected to be forthcoming in the next few years as further refinements are made. No attempt has been made to give papers dealing specifically with glow and arc discharges since these discharge forms are rather outside the scope of the present selection. However, their development must be seen as part of the overall development of the breakdown process (see for example the paper by Meyer and Cavanor, cited by Haydon in paper 14, which studied the glow-to-arc transition).

The special forms or conditions of breakdown included here are breakdown at low values of the product pd (pressure x gap distance), corona breakdown, breakdown in long discharge gaps of non-uniform fields, lightning discharges, and laser-induced breakdown. This does not, of course, cover all the available possibilities. For example, papers on breakdown at high pressures have not been selected, neither have any papers on so-called vacuum breakdown. The paper on breakdown at low values of (pd) is the earliest of the fairly limited number published. It was established by Paschen nearly eighty years ago that the graph of breakdown potential V_s as a function of the product of the gas pressure and gap length possesses a minimum value, $(V_s)_{min}$, at a particular value, $(pd_s)_{min}$, of the independent variable. Since that time most studies of breakdown have been carried out at values of pd greater than $(pd)_{min}$. The range of values of (pd) less than $(pd)_{min}$ is of considerable interest, however, particularly in view of the rapid changes in V_s produced for very small percentage changes in (pd). The theoretical study of breakdown at low (pd) requires some compromise between Townsend theory and some suitable extension of that based on the physics of single collisions and is not well developed.

The paper on corona breakdown is the second case where a review paper has been selected. There is such a wealth of published experimental data that to select any single paper is impossible. The interested reader is strongly advised to refer to the comprehensive summary given by Professor Loeb in *Electrical coronas* for a thorough introduction to the field and to the relevant chapters of Professor Nasser's recent book *Fundamentals of Gaseous Ionization and Plasma Electronics*. The subject of corona breakdown is, of course, closely linked to that of breakdown in the non-uniform field geometries of long gaps such as those described in the paper by Stekolnikov and Shkilyov which in turn is closely linked to papers 21, 22 and 23 on studies of lightning discharges.

The final two papers (24 and 25) are among the earliest published on gaseous electrical breakdown induced through the action of high-power laser radiation. This is a rapidly developing field of study in spite of the experimental and theoretical difficulties involved. The breakdown mechanisms involved are closely related to those which operate in so-called micro-wave discharges. Laser beams are increasingly utilised in constructing triggered spark-gaps and the devices have been shown to be capable of high-speed operation with the minimum of delay and jitter.

J. A. REES

Introduction to Paper 1 (Townsend, 'The theory of ionization of gases by collision')

It is appropriate to begin a selection of papers on the electrical breakdown of gases with brief extracts from a book by J. S. Townsend, since it was Townsend, assisted by his students, who carried out a large proportion of the early investigations into the phenomena accompanying the electrical breakdown of gases, and developed the theory of the breakdown process which bears his name. The extracts given here summarise some of the pioneering experiments carried out by Townsend, emphasise the importance of the parameter E/p, and introduce the parameter α which became known as 'the Townsend ionisation coefficient' or 'the Townsend primary coefficient'.

In attempting to explain why at large values of l the electron current in a parallel-plate discharge gap increased faster than that predicted by the simple exponential law $n = n_0 \exp{(\alpha l)}$, Townsend considered the ionisation produced in the gas by positive ions. The equation derived was

$$n = \frac{n_0(\alpha - \beta) \exp\{(\alpha - \beta)l\}}{\alpha - \beta \exp\{(\alpha - \beta)l\}} \qquad (1)$$

It is now known that ionisation by positive ions is inadequate to account for observed data. However, an equation of similar form, *viz.*

$$n = \frac{n_0 \exp{(\alpha l)}}{1 - (\omega/\alpha) \{\exp{(\alpha l)} - 1\}} \qquad (2)$$

can be derived where the coefficient ω is the so-called 'generalised secondary ionisation coefficient'. (To see that the two equations are analytically of the same form, substitute $(\alpha - \beta) = \alpha'$ and $\beta = \omega$ in eq. 1). ω includes the β process considered by Townsend but also other processes such as the action of positive ions, photons and metastable atoms at the cathode. Different secondary processes may dominate under different experimental conditions. Full details are given in the references.

A *breakdown criterion* corresponding to that given by Townsend ($\alpha - \beta \exp\{(\alpha - \beta)l\} = 0$) can be deduced and is $1 - (\omega/\alpha) \{\exp{(\alpha l)} - 1\} = 0$.

Further Reading

Llewellyn Jones, F., *Ionization and Breakdown in Gases*, Methuen, London, (1966), pp. 46-61.

Morgan, C. Grey, *Handbook of Vacuum Physics*, Beck, A. H. W. ed., Volume 2, Part 1, (1965), pp. 66-9.

1

Extracts from *The theory of ionization of gases by collision*

J. S. TOWNSEND*

Ionization by negative ions†

Variation of current with electric force

The process of ionization by collision between ions and molecules of a gas may be examined by investigating the currents between parallel plate electrodes when ultra-violet light falls on the negative electrode or when the gas is ionized by Röntgen rays. If the gas is at a high pressure, the current increases with the electric force and attains a maximum value, which is not exceeded unless very large forces are used. It is possible, however, by reducing the pressure of the gas, to make the ions travel with sufficient velocity to generate others by collisions with molecules, even when the potential differences employed are small, and thus with a few hundred volts to obtain large increases in the current.[1] The curve, *Figure 1*, showing the connection between the current and electric force in a gas at low pressure, may be taken as illustrating this effect.

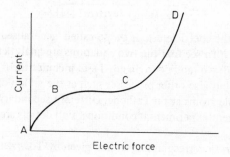

Figure 1.

* Reproduced from *The Theory of Ionization of Gases by Collision* by J. S. Townsend, Constable & Co. Ltd., London, (1910) by the permission of the publisher.
    † *Editor's note:* The 'negative ions' referred to in this paper are electrons.

In the first stage, AB, the current between the plates increases with the electro-motive force. The rate of increase diminishes as the force increases, and the current tends to attain a maximum value.

In the second stage, BC, the current remains practically constant and shows only small variations for large changes in the force. If the ions are produced by the action of Röntgen rays or Becquerel rays, the constant value is attained when the force is sufficiently great to collect all the positive and negative ions on the electrodes, but before this value is reached an appreciable number of the ions is lost by recombination. Again, the ions may be produced by the action of ultra-violet light on the negative electrode. In this case, if the force is too small, some of the ions do not reach the positive electrode, but diffuse through the gas to the negative electrode.

In the third stage, CD, when the force is still further increased, there is a large increase in the conductivity. This can be explained on the hypothesis that new ions are generated by collisions, at first practically by negative ions alone, but as the force increases and the sparking potential is approached, the positive ions also acquire the property of producing others to an appreciable extent.

Variation of current with distance between the electrodes when the force is constant

In the earlier experiments which were made to test the theory the initial ionization was produced by the action of Röntgen rays. The simplest conditions, how-ever, are realized when the initial ionization consists of negative ions set free from a metal surface by a beam of ultra-violet light.

When the light falls on a metal plate a number n_0 of negative ions are set free which can be made to travel various distances through a gas under any required force to a parallel plate positively charged. If no new ions are produced by collisions the number reaching the positive plate will be n_0, and the current will be independent of the distance between the plates. If, however, the ions produce others by collisions with the molecules of the gas between the plates, the number reaching the positive plate will increase and will depend on the distance between the plates. In fact, if each negative ion set free from the metal plate produces α new negative ions in going through a centimetre of the gas, and if the new ions produced in the gas have exactly the same property of generating others by collisions, then the number that arrive at the positive plate will be $n_0 \exp(\alpha l)$ where l is the distance between the plates. For let n be number of ions produced in a layer of thickness, x, measured from the negative electrode, the number n including the original n_0 ions. In passing through a path of length dx these ions produce $n\alpha dx$ new ions by collisions, so that $dn = n\alpha dx$. This equation on integration gives $log\, n = \alpha x + constant$, or $n = n_0 \exp(\alpha x)$ since n_0 is the value of n corresponding to $x = 0$.

The quantity α depends only on the electric force and pressure of the gas, and if these are constant the charges n_1, n_2, etc., acquired by the positive plate for different distances l_1, l_2, etc., between the plates will be

$$n_1 = n_0 \exp(\alpha l_1); n_2 = n_0 \exp(\alpha l_2); \text{etc.}$$

Hence for equal increments of the distance the ratios of the successive charges will be the same, *viz.*

$$\frac{n_2}{n_1} = \frac{n_3}{n_2} = \text{etc.} = \exp\{\alpha(l_2 - l_1)\}.$$

The conditions specified above are easily realized in practice, and experiments show that this simple exponential law for the increase of the current with the distance l is accurately true for small distances between the plates; for larger distances the conductivity rises more rapidly owing, as will be explained later, to the effect of positive ions.

Representation of the values of α by a single curve

In order to examine the values of α for different forces and pressures, it is convenient to present the experimental results in a special way by means of a curve for each gas. The method leads to great simplification as the value of α corresponding to any force and pressure may be obtained immediately from the curve. If α be determined for a force E and pressure p, and the points whose co-ordinates are α/p and E/p be marked on a diagram, it will be found that they all lie on the same curve, or if the experiments be examined it will be noticed that if α' corresponds to a force E' and a pressure p', then at a pressure zp', the value of α will be $z\alpha'$ when the force is zE', z being any multiplier.

In the later experiments the pressures were chosen so as to show this by inspection.[2] Thus, with air at a pressure of 1 millimetre and a force of 350 volts per centimetre, $\alpha = 5\cdot25$; at 2 millimetres pressure and with a force of 700 volts per centimetre, $\alpha = 10\cdot5$. Another example may be taken from the experiments with hydrogen:

$$p = 8 \text{ mm}, E = 1050 \text{ V/cm}, \alpha = 14\cdot8$$
$$p = 4 \text{ mm}, E = 525 \text{ V/cm}, \alpha = 7\cdot4$$
$$p = 2 \text{ mm}, E = 262 \text{ V/cm}, \alpha = 3\cdot7$$

The results of all the experiments for any one gas can therefore be recorded by a table of values of α/p and E/p or by means of a curve[3] representing one of these ratios as a function of the other, $\alpha/p = f(E/p)$.

Agreement between the experimental results and the equation $\alpha/p = f(E/p)$

It is easy to see that the theory requires that this relation should exist between the variables α, E, and p.

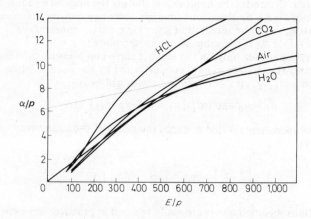

Figure 6. E in volts per centimetre, p in millimetres of mercury.

In passing through a centimetre in a gas an ion traverses free paths of various lengths between the collisions. The chance of producing a new ion by collision will depend on the velocity at impact, and this is determined by the force E and the length of the path which is terminated by the collision. The lengths of the free paths are inversely proportional to the pressure, so that if the pressure is increased from p to zp all the free paths will be reduced to $1/z$ of their original value. If the force E remained unaltered, the velocities on collision would be reduced, but if the force is increased to zE the velocities will be restored to their original values and the number of ions arising from a given number of collisions will be the same as before. Since the total number of collisions per centimetre is increased in the same ratio as the pressure, the value of α will be increased to $z\alpha$, when E and p become zE and zp respectively. Hence, the three variables must be connected by an equation of the form $\alpha/p = f(E/p)$. The experiments confirm this result very accurately for all the gases and so afford an important verification of the theory.

The curves representing the values of α/p corresponding to some of the larger values of E/p are given in *Figure 6*.

Ionization by positive ions

Conductivity between parallel plates when positive and negative ions generate others by collisions

The effects produced by the motion of positive ions may be deduced from determinations of the currents that pass between parallel plates while ultra-violet light is falling on the negative electrode, if the forces and the distances between the plates are sufficiently large.

It has already been shown that when E/p and l are small, and l is varied whilst E and p are kept constant, the number of negative ions reaching the positive electrode is $n_0 \exp(\alpha l)$, where α is a constant depending on E and p. For large values of E/p and l the number of negative ions reaching the positive electrode is greater than $n_0 \exp(\alpha l)$, showing that some other form of ionization has come into play. This stage is attained even when the potential between the plates is much less than that required to produce a spark. It will be seen from the following investigations that all the features of the new process of ionization can be explained on the supposition that it arises from the action of the positive ions.

Further, it is obvious that if both positive and negative ions produce others in sufficient numbers, a current would be obtained which would last after the supply from the negative electrode was cut off, and a continuous discharge would ensue. The investigations show how the potential required to produce a continuous discharge may be found on the assumption that the whole ionization is produced by collisions of positive and negative ions in a gas, and, as will be seen, there is a very accurate agreement over a large range of pressures between the potentials thus calculated and the sparking potentials determined experimentally.

In making an investigation of the currents which would be produced between parallel plates when both positive and negative ions generate others by collisions, the results of the experiments may be anticipated, and it may be assumed that the positive and negative ions produced by the impact of a positive ion with a neutral molecule are identical with the positive and negative ions produced by the impact of a negative ion. In applying the theory, therefore, it is necessary only to consider one kind of positive ion and one kind of negative ion in each gas, but the positive ions, unlike the negative ions, are most probably different in different gases.

If a number n_0 of negative ions start from the negative plate and move through a distance l from the positive plate they will generate others in the gas, and a number $n_0 \exp(\alpha l)$ will reach the positive electrode. Thus $n_0 \{\exp(\alpha l) - 1\}$ positive ions are produced in the gas and move in the opposite direction. When these also ionize the gas the total number n of negative ions that reach the positive electrode exceeds $n_0 \exp(\alpha l)$.

Of the number $n - n_0$ of each kind generated in the gas, let r be produced in the layer of gas between the negative plate and a parallel plane at a distance x, and let r' be produced in the layer between this plane and the positive electrode

$$\text{Then } n = n_0 + r + r'$$

Let α be the number of ions produced by a negative ion in moving through one centimetre of the gas.

Let β be the corresponding number for a positive ion.

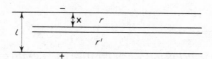

The number of ions dr generated between the two planes at distances x and $x + dx$ is given by the expression $dr = \alpha(n_0 + r)dx + \beta r'dx$, since the number of negative ions travelling through the length dx is $n_0 + r$, and these produce $\alpha(n_0 + r)dx$, and the number of positive ions passing through the same element of length dx is r', and they produce $\beta r'dx$. Substituting $n - n_0 - r$ for r' the equation for r becomes

$$\frac{dr}{dx} = (\alpha - \beta)(n_0 + r) + \beta n,$$

which on integration gives

$$n_0 + r = -\frac{\beta n}{\alpha - \beta} + c\exp\{(\alpha - \beta)x\}.$$

The constant c is determined by the condition $r = 0$ when $x = 0$.
Hence the value of r is given by the equation

$$n_0 + r = -\frac{\beta n}{\alpha - \beta} + \left(n_0 + \frac{\beta n}{\alpha - \beta}\right)\exp\{(\alpha - \beta)x,\}$$

and the value of n in terms of the other constants is obtained by substituting for r its value $n - n_0$ when $x = l$, l being the distance between the plates. Thus

$$n = n_0 \frac{(\alpha - \beta)\exp\{(\alpha - \beta)l\}}{\alpha - \beta\exp\{(\alpha - \beta)l\}}.$$

This gives the number of negative ions n arriving at the positive electrode when n_0 negative ions start from the negative electrode, and both positive and negative ions generate others by collisions with molecules of the gas.

It will be noticed that when the distance between the plates l has a certain value S given by the equation

$$\alpha - \beta \exp\{(\alpha - \beta)S\} = 0, \quad \text{or} \quad S = \frac{1}{\alpha - \beta} \log\left(\frac{\alpha}{\beta}\right),$$

the denominator of the fraction in the expression for n becomes zero, so that n becomes infinite. This shows that a current would continue to flow indefinitely after the supply of the negative ions n_0 from the surface of the negative electrode ceases. The importance of this conclusion in connection with sparking potentials will be considered later.

For distances between the plates shorter than S the denominator of the fraction expressing n is positive, and the current becomes zero after the light ceases to act on the electrode. The values of n are then finite, but greater than the number of ions $n_0 \exp(\alpha l)$ which would reach the positive electrode if the negative ions alone produced others by collisions.

Condition for sparking in a uniform electric field

The sparking potential between parallel plate electrodes can be deduced immediately from the formula

$$n = n_0 \frac{(\alpha - \beta) \exp\{(\alpha - \beta)l\}}{\alpha - \beta \exp\{(\alpha - \beta)l\}},$$

which represents the number of ions n reaching the positive electrode when n_0 negative ions are started by ultra-violet light from the negative electrode. The quantity n increases as the distance l between the plates increases, the force E being maintained at a constant value, and when $\beta \exp(\alpha - \beta)l$ becomes equal to α the denominator of the above fraction vanishes, and n becomes infinite. A current will therefore continue to flow for an indefinite time after the light ceases to act when the distance between the plates attains the value S given by the equation

$$\alpha - \beta \exp\{(\alpha - \beta)S\} = 0.$$

Thus the sparking potential SE is known for the particular distance S and pressure p of the gas.

References

1. Townsend, J. S., *Nature*, **62**, 340, (1900).
2. Many examples may be found in the researches published in *Philosophical Magazine*, November 1903, and December 1904.
3. This result was first obtained from the experiments with Röntgen rays, see *Philosophical Magazine*, February 1901.

Introduction to Paper 2 (Rogowski, 'Impulse potential and breakdown in gases')

The paper from which this extract is taken is typical of a number by Rogowski and others, in which oscillographic techniques were used to examine the current flowing in discharge gaps after the application of voltages greater than or equal to the breakdown voltage. The experiments were in fact among the earliest investigations of 'time lags', i.e. of the time that elapses between the application of a voltage greater than the breakdown voltage and the time at which breakdown may be said to have occurred.

The general conclusion from the work of Rogowski and others was that the Townsend theory, as outlined in paper 1, was unable to account for the short time lags observed, the rapid rate of rise of the current in the discharge, or the large currents that were seen to accompany breakdown. This was particularly the case when the secondary ionisation was assumed to be due to ionisation of the gas by positive ions, since the time lags were orders of magnitude shorter than the time taken by positive ions to cross the discharge gap.

Later experimental studies of time lags have, of course, made use of considerably more sophisticated techniques. Some of these later results and conclusions are described in papers 7, 8, 9 and 10.

2

Extract from *Impulse potential and breakdown in gases*

W. ROGOWSKI*

Previous results

In an earlier paper[1] I was able to show that the well-known Townsend theory of
spark discharges, which has hitherto been taken as generally valid, runs into great
difficulties when it is applied to the results of tests on the breakdown of air under
impulse potentials. For instance, Townsend assumes that, when the potential is
high enough, both an electron avalanche and a positive-ion avalanche will appear
in the gas. These move in opposite directions and contribute to each other's
amplification. If these processes are not interrupted, large currents flow and finally
breakdown occurs. The order of magnitude of the velocity of both types of
avalanche can be estimated. We may assume the air to be at atmospheric pressure
and to have a breakdown field strength of about 30 000 V/cm (spark length 1 cm).
As I showed earlier, the electron avalanche has a velocity of about 10^7–10^8 cm/s,
and the positive-ion avalanche a velocity of about 10^5 cm/s. If we use these values,
it follows that we shall only get significant currents when the ordinary static breakdown potential can act on the spark gap for well over 10^{-5} s. Hence to get breakdown, we would expect to obtain higher values for the spark current from about
10^{-4} s onwards. However, all earlier experiments showed that this increase actually
begins at about 10^{-6} s, and perhaps even at 10^{-7} s. In fact, breakdown takes place
10^{-2} or 10^{-3} s earlier than would be predicted by the Townsend theory. Herein
lies the difficulty. Either the calculation of the avalanche velocities,[†] the experiments,
or the principles of the Townsend theory must be at fault.

Reliability of the velocities

The point which really brings the uncertainties to light is the low velocity of the
positive ions, about 10^5 cm/s at 30 000 V/cm. Its calculation is based on having a
single species of ion and the ordinary value for the mean free path. However recent
measurements by Dempster have revealed the possibility that these ions can also
attain greater mean free paths. He has obtained nine times the usual value for

* Reproduced from *Archiv. für Elektrotechnik* **20**, 99, (1928) by permission of Springer-Verlag,
Berlin.

† Other people besides myself have questioned the accuracy of the Townsend theory. See
Schumann, *Z., Tech. Phys.* 620, (1926); Seeliger, *Gasentladungen*, Ambr. Barth, Leipzig 170,
(1927); Holm, *Arch. Elektrotech.* 18, 80, (1927).

hydrogen ions in helium. Of course he did obtain the usual value for helium ions in helium. Nothing seems to be known about the behaviour of oxygen or nitrogen ions. In spite of this, our values for the velocity as given above can be looked on as still valid order of magnitude values. This means that it only increases with the root of the mean free path.[1] If we want to explain the extreme suddenness of breakdown in terms of the Townsend theory, we have to multiply the positive-ion velocities by 100, or even 1 000, getting path lengths which are 10^4 to 10^6 times higher than usual. At this stage it is difficult to assume that they can reach such a value. Finally, as will be seen a little later from the spark oscillograms, we have to explain not only the high velocities but also the sudden appearance of strong currents. These are however, associated with multiple ionization and loss of energy and retardation of the ions.*

The velocity given earlier for the electron avalanche of about 10^7–10^8 cm/s involves no particular difficulties for the theory. Experiments by Ramsauer and Bruche have given us detailed information on the features of the free paths of electrons. The measured values for the free path lengths for cold electrons in nitrogen are about half those for gas-kinetic electrons. No particular difference was found for oxygen; thus the velocity of the electron avalanche seems to have been over-estimated rather than under-estimated.

Reliability of the experiments

As I pointed out earlier, the weakness of the earlier experiments used to test the theory was that the instantaneous voltage at the spark gap was not measured but calculated. At that time, I proposed measurements with the cathode-ray oscilloscope. They are described in papers by myself[2] and by my co-workers Flegler and Tamm.[3] Our photographs not only confirmed the discrepancy between theory and experiment, but intensified it. Our results can be summed up as follows; if the static breakdown voltage is suddenly applied to an (irradiated) spark gap, breakdown will follow fairly regularly after 10^{-7} s. If the breakdown voltage is raised by only 30%, the spark time lag shrinks from 10^{-7} to about 10^{-8} s (see *Figures 1* and *2*).

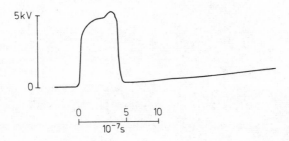

Figure 1. Breakdown of a gap between parallel plates on application of a static breakdown voltage.

* It will be noted from the oscillograms of *Figure 1* and *Figure 2* that at a 30% overvoltage the spark time lag is reduced from about 4×10^{-7} to about 10^{-8} s. Since the velocity of the avalanche increases in proportion to the square root of field strength, the velocity is reduced by about 10^4.

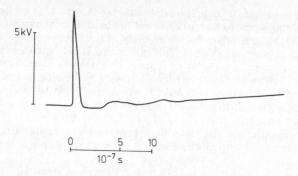

Figure 2. Breakdown of a gap between parallel plates on
application of an overvoltage.

Conclusion

We can now say with absolute certainty that the experimental results have been confirmed and that the values used for the velocities in the comparison are accurate enough for the purpose. Thus the inescapable conclusion is that, though the Townsend theory cannot be entirely brushed aside, it does contain elements that are in contradiction to the facts. The idea of two avalanches moving in opposite directions through the entire discharge space cannot be considered to be accurate today.

References

1. Rogowski, W., *Arch. Elekt.*, **16**, 496, (1926).
2. Rogowski, W., *Arch. Elekt.*, **18**, 504, (1927).
3. Tamm, R., *Arch. Elekt.*, **19**, 235, (1928).

Introduction to Paper 3 (Raether, 'The development of the electron avalanche in a spark channel (from observations in a cloud chamber'))

This paper is one of the early ones in a series on the important studies carried out by Raether and his co-workers using a cloud chamber as a technique for studying electrical discharges. As described in the paper, the technique enabled the observation of the development in space and time of a low current pre-breakdown electron avalanche whose initial spatial growth was described by the Townsend equation $n = n_0 \exp(\alpha x)$ into a high current spark. The particular paper given here set out to establish whether the track of an avalanche initiated by a single electron from the cathode represents the track of the spark ultimately produced and if so in what way the transition occurs. The investigation used impulse discharges throughout rather than static d.c. discharges, the impulse voltages used being 10 to 20% greater than the static breakdown potential, but the results were generalised to apply also to the static case.

Raether interprets the results of the experiment as showing that at a critical amplification ($\alpha x \approx 20$, where α = ionisation coefficient and x = avalanche length) the electron avalanche becomes unstable and a spark track starts developing towards the anode, at a high velocity $\approx 8 \times 10^7$ cm/s, from the head of the avalanche. A little later a track starts developing towards the cathode also. The high speed of the development of the tracks is attributed to the high space-charge fields at the tips of tracks along with the production of gas-ionising radiation. If the cathode-directed track, or 'kanal', reaches the cathode, a narrow track bridges the two electrodes; this is faintly luminous at first but the current increases rapidly and the light output reaches that of a 'spark'.

The results of this and similar papers appeared to be difficult to explain on the basis of the ideas set out by Townsend and others in papers such as paper 1 and were one reason for the attempts made by Raether, Loeb, Meek and others to formulate alternative mechanisms for the development of spark breakdown. The following papers 4 and 5 describe these alternative mechanisms.

Further Reading
Raether, H., *Electron Avalanches and Breakdown in Gases* Butterworths, London
(1964) particularly sections 2.1 and 5.2.1.

3

Extracts from *The development of the electron avalanche in a spark channel (from observations in a cloud chamber)*

H. RAETHER*

Aim of the research

The aim of the present research is to investigate in more detail the development of the electron avalanche, which has so far been studied within the region of exponential amplification.[1] This is linked to the question of whether the avalanche track of the *one* electron represents the future spark track, and if so in what way the transition from the normal avalanche to the spark track takes place.

Introduction

The use of the cloud chamber makes it possible to study the initial processes of collisional ionization, which take the form of an 'electron avalanche'. The development of the avalanche in time and space can be followed and explained in almost quantitative terms from the movement of the electrons.

Townsend has established from studies of the current-distance characteristics under static conditions that, at high amplifications, i.e. large values of $\exp(\alpha x)$, new processes become apparent as well as collisional ionization, giving rise to a greater increase in current than $\exp(\alpha x)$. These processes can be: collisional ionization by the positive ions; release of electrons from the cathode by positive ions or ultraviolet-light quanta; photo-ionization by radiation that ionizes the gas; or space-charge effects. If the value of $\exp(\alpha d)$ is great enough, either because a suitable value has been chosen for the distance d between the electrodes or because α is enlarged by the use of a high field strength, these processes in the discharge region can produce so many electrons that the discharge itself is maintained by these 'secondary electrons', or can be converted into a high-current discharge such as a spark.

Now we did not examine any static discharges in the cloud chamber, but only those due to impulse voltages. We know however that these voltage impulses not only produce 'electron avalanches' but can also develop them to the point of a 'spark' (impulse breakdown). In any event the voltage must be higher than the static breakdown voltage (overvoltage) to produce a spark if the duration of the impulse voltage is shorter than the time of flight of an ion from anode to cathode.

* Reproduced from *Zeitschrift für Physik,* **112,** 464, (1939) by permission of Springer-Verlag, Heidelberg.

Because of the non-static conditions, we come up against the same question again: which of the processes mentioned earlier leads with increasing exp (αd) to the development of the normal electron avalanche into the spark channel, and in what way does the transition take place?

To answer this question with the aid of the cloud chamber, we follow the appearance of an individual electron avalanche as the amplification increases. However, present-day theoretical treatments of the problem of breakdown usually take the view that a spark can only be produced by the interaction of several avalanches, and that the space charge of the first avalanche can alter the ionization conditions for the subsequent avalanches. If this were indeed the case there would be no point in trying to make observations in the cloud chamber, since a sequence of avalanches in the discharge zone would make it impossible to follow what was happening. However, as the following experiments have shown, even at a considerable over-voltage the spark track can develop from a *single* electron avalanche, so that we can answer our previous question by following the development of *one* avalanche at increasing collisional amplifications.

There are two possible ways of doing this: the collisional amplification can be increased by:

(a) using a greater field strength (larger α), with a constant collision time, i.e. a fixed avalanche length;

(b) prolonging the collision time (greater x) with a constant field strength, i.e. fixed α.

Both methods will be described here. The experimental method was the same as that used in reference 1. It consisted of a pulse unit that produced approximately rectangular voltage pulses of up to 3×10^{-7} s duration. The space of 3·6 cm between two plane electrodes, to which the impulse voltage was applied, was used as a cloud chamber so that the ionization processes taking place in the space between the electrodes could be followed. The avalanches were initiated by photoelectrons, which were released from the cathode once the voltage applied to the cloud-chamber electrodes had reached its highest value. This eliminated the scatter brought about by waiting for electrons. For more details, see reference 1, page 93.

Part A: Results
1. Observations made with normal expansion conditions in the cloud chamber (air at 273 torr, 20°C)

(a) *Development of the electron avalanche when voltage is increased and pulse duration is held constant*

If the voltage to the electrodes is raised and the pulse duration is kept constant, it is observed that: at 11·4 kV/cm a normal electron avalanche about 1·9 cm in length develops, more or less in the manner shown by the arrows in *Figure 1(b)*, (see also the diagrams in reference 1). At 12 kV/cm there is a sudden vigorous expansion of the head of the avalanche, whilst the foot retains its shape (already described in *Verh. d. Phys. Ges.* 18, 54, (1937)). The appearance of the avalanche in this stage is shown by *Figures 1(a)* and *1(b)*. At about 12·3 k V/cm the avalanche has developed into a wide track (see *Figure 1(c)*).

When viewed in the dark at about 12 kV/cm the head of the avalanche is seen to emit a very weak diffuse light; at about 12·3 kV/cm a thin thread of light is seen from electrode to electrode. A further slight increase in voltage up to 12·6 kV/cm causes this thread of light to be intensified rapidly to an intense blue track. It is

a   ↑   ↑   b   c

Figure 1. Development of an electron avalanche indicated by
the arrows in *b*, in a continuous channel (with normal
expansion ratio in the chamber). In the sequence *a*, *b* and *c*
either the voltage is raised and the pulse time kept constant, or
vice versa. *b* represents a more advanced stage than *a*, as
revealed by the appearance of a track developing towards the
anode. *c* shows complete development into a side-to-side
channel (air, 273 torr); top = anode; bottom = cathode; as also
for *Figures 2, 3* and *4*.

worth noting that a particularly bright patch can often be seen at the top of the
avalanche head in the track. The avalanche develops in the same way whether its
length is longer or shorter. The values for an avalanche 3 cm in length may be given
by way of example:

10·9 kV/cm   normal avalanche
11·3 kV/cm   intensive swelling of
the head of the avalanche;
further small increase in
voltage gives rise to a
track from side to side.

If we calculate the value of αd (α = electron collisional ionization factor, d = length
of avalanche) for the field strength at which expansion of the head begins, it is found
to be about 20, independent of what length of avalanche is being used.

As can be seen from *Figure 1(b)* in particular, a track develops towards the anode,
independent of the thickening of the avalanche head: it will be shown later that this
track moves at a high velocity towards the anode. It is difficult to follow the details
of this process under the experimental conditions described, since the vigorous
formation of ions contaminates the chamber and prevents the use of a rapid series
of expansions. For more details see section **2**.

(b) *The development of the electron avalanche when the duration* of the voltage
pulse is extended while the voltage is kept constant.*

The avalanche develops strictly according to this pattern:

12 kV/cm:

Length of cable:	22 m	normal avalanche	(*Figure 1(b)*↑)
Length of cable:	25 m	expanded head	(*Figure 1(a)*)
Length of cable:	27 m	expanded head	(*Figure 1(b)*)
Length of cable:	29 m	broad side-to-side track (thin luminous thread)	(*Figure 1(c)*)
Length of cable:	30–31 m	strongly luminous spark track	

* The durations of the pulses are given in terms of the length of cable used to produce the
pulses. The pulse duration is taken as approximately

$$\frac{2 \times \text{cable length (in metres)}}{3 \times 10^8} \text{ seconds.}$$

Observations of avalanche development with prolongation of the pulse time make it possible to estimate their velocity. Since the electron avalanche develops within $(8/3) \times 10^{-8}$ s $(2 \times 4$ m) into a complete channel, i.e. bridges at least the distance of 3·6 cm between the electrodes in this time, we get a rate of development of about $1·3 \times 10^8$ cm/s. With the normal development (velocity $u = 1·25 \times 10^7$ cm/s) (reference 1, p. 100) the avalanche would only become about 3 mm longer in this time.

Here too (as in *Figure 1(a)*) the avalanche head begins to thicken at $\alpha x \approx 20$, so that we can conclude that at this critical amplification the transition to the spark track is accompanied by a process with a high rate of development.

This result was confirmed by observation of the luminous phenomena: at about 12 kV/cm and 25 m cable length the head of the avalanche appears faintly luminous when viewed in a totally darkened room, until a weak thread of light travels from the avalanche head to the anode. Thin continuous channels are already visible at 29 m. The following describes a series of experiments which revealed how many times during ten voltage pulses illumination of the avalanche head (L) or a continuous luminous thread (K) was observed:

Length of cable (m)	25	27	29
L	10	5	1
K	0	5	9

(On the whole, the continuous luminous tracks appearing at 27 m were feebler than those appearing at 29 m.) Thus we get values of more than 10^8 cm/s for the rate of development of the continuous channels. A rapid transition to the channel was also observed at pressures of 80 torr.

The development of the electron avalanche, from narrow avalanche track to expanded head to continuous channel takes place in the same way in hydrogen.

Another method must be used to get details of the method of transition from the normal electron avalanche to the continuous channel, since the strong cloud formation round the traces makes it difficult to follow their development.

2. Observations with sub-normal expansion conditions in the cloud chamber (Air: 20°C)

Preliminary remarks

It has been established by several workers[2,3] that cloud traces of electric discharges can be obtained at lower expansion ratios than normal. It is assumed that at sub-normal expansions only the places of maximum ion concentration can act as condensation centres.

The appearance of gas discharge traces at sub-normal expansion has so far been studied in the case of emission from a point, for which the current cannot be controlled, and also for the case of well-advanced discharges in homogeneous fields, so that many factors can be adduced to account for the observed phenomenon.

However it is also found to occur in an electron avalanche whose ion density can be controlled in a regular way. It has been observed that: with a constant pulse time (i.e. a stable avalanche length of 1·9 cm) and the same initial pressure (320 torr, 20°C) the field strength E/p can be measured as a function of the expansion ratio ϵ, E/p being that at which the first droplets become visible in the chamber.

e	1·18	1·16	1·15	1·14	1·13
E/p (0°C)	36·8	37·6	41·4	45·2	45·3
αx	3·4	3·9	9·4	16	16·5

The first two values of αx correspond to my earlier observations on the onset of collisional ionization (reference 1, p. 98), the values for the lower expansions correspond to a possible condensation state that only becomes active at high amplifications. This possibility is probably afforded by doubly ionized molecules that need less supersaturation since the double charge of the vapour pressure on the surface of a small sphere corresponds more closely to that on a plane surface than a single charge. Since doubly ionized molecules are produced preferentially at points where there is a high concentration of carriers (electrons), these regions of dense ions will become visible at low expansion ratios.

This fact was used to follow the development of the avalanche more exactly in the following way: if the pulse time is fixed so that a normal avalanche crossing half the electrode gap is produced, the expansion ratio is lowered but E/p is kept constant. However all that is visible as a rule is the ion-dense avalanche head in the middle of the discharge zone and its further development can be followed as the voltage is increased or the pulse time prolonged. For the following observations the expansion ratio was lowered from 1·17 (1:3 mixture of alcohol and water) to 1·13, and a final pressure of 260 torr (20°C) was used.

(a) Pulse time constant, voltage increased

Under the conditions given and with a 20 m cable, we observe a cloud track a few millimetres long about 1·3 cm from the cathode at 11·8 kV/cm (cf. *Figure 2(a)*). (The series of photographs in *Figure 2* were taken with a pulse time 8 m shorter ($\approx 2\cdot7 \times 18^{-8}$ s), and really relate to experiments that will be described later, but the development would be exactly the same for the longer pulse time described here.) The distance of the track from the anode varies, but its distance from the cathode remains the same. As the voltage increases the track develops towards the anode (*Figures 2(b)* and *2(c)*) until at about 12·2 kV/cm there is a track extending across the entire discharge region (*Figures 2(d)* and *2(e)*). We get the result that the transition from avalanche to track takes place in such a way that first the gap between the avalanche head and the anode is bridged by an 'anode-side' track, and then the gap to the cathode is bridged by a 'cathode-side' track.

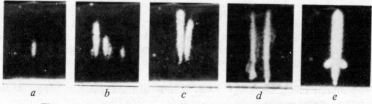

$\qquad a \qquad\qquad b \qquad\qquad c \qquad\qquad d \qquad\qquad e$

Figure 2. Development of an electron avalanche in the side-to-side channel at a subnormal expansion ratio. The expansion ratio has been lowered so far that only the region of the head of an electron avalanche is visible (*a*). The voltage in the discharge zone is raised from *a* to *d* (see text); *d* and *e* are taken at the same voltage. The photos show that a track first develops from the avalanche head to the anode and then the track to the cathode develops.

The length of the anode-side track can be varied with the voltage within narrow limits, but the cathode-side track is either formed or not, i.e. its development is much more dependent on the voltage than that of the anode-side track. In only a few cases did the voltage at the electrodes fall low enough to prevent the growth of the cathode-side track. In these cases branches were seen developing from the avalanche head towards the cathode, as shown in *Figure 3(b)*. In most cases however the track went direct from electrode to electrode. Protuberances could be seen on both sides directed towards the cathode (*Figure 2(e)*), always at the level of the first track to become visible. (*Figure 2(a)*, avalanche head.) The starting point for these protuberances is obviously identical with the lighter patches in the avalanche head mentioned previously. If the room is darkened, both the halfway and the full tracks are visible as faint blue lights.

(b) Voltage constant, pulse time increased

The electron avalanche goes through the same stages when the pulse-time is extended, so that we can get an idea of the rate at which the processes develop.

If we start with the same conditions as before and increase the cable by 3 m, a complete track from the core of the avalanche to the anode is almost always formed. In this case a distance of about $1·4$ cm has to be traversed, so that the mean rate of development of the anode track works out at about 7×10^7 cm/s. If we extend the cable by another $1·5$ m, we get the side-to-side track. This means that a distance of about $1·2$ cm has been traversed, assuming the development to have begun in the avalanche head, and so we get a minimum velocity of $1·2 \times 10^8$ cm/s. Development towards the cathode is thus quicker than that towards the anode. These numerical values were obtained by evaluating a number of films.

The following experiments were then undertaken for the complete explanation of this picture:

1. Observations with short pulse times

The core of the avalanche was displaced towards the cathode by shortening the pulse time (the distance was about 1 cm) and attention was directed towards the development of the anode-side track. Again, the interval between the first appearance of the avalanche core ($12·7$ kV/cm) and the appearance of a continuous track ($13·2$ kV/cm) was small. It was again confirmed that the anode-side track formed before the core-to-cathode link. *Figure 2* shows the various stages. The duration of each of these phases is as follows: at a field strength of about $12·9$ kV/cm the cloud trace has moved about $1·6$ cm towards the anode if the cable is extended by 3 m. Repeated measurements of this type, visual as well as photographic, gave values between $7–9 \times 10^7$ cm/s for the mean rate of development towards the anode, similar to what was found under section 1(b).

2. Observations with longer pulse times

To get a more exact idea of the development of the track towards the cathode, the avalanche core was displaced towards the anode by increasing the pulse time so that at about $2·2$ cm from the cathode the avalanche core appeared at about $10·9$ kV/cm (*Figure 3(a)*). At about $11·4$ kV/cm the side-to-side kanal had formed (*Figure 3(c)*). As a result of the larger distance between the core and the cathode it was often possible as in *Figure 2(a)* to get only half-developed tracks (*Figure 3(b)*), and thus it was made obvious that the cathode-side track develops from the avalanche head in the direction of the cathode. To measure the duration of this return growth, the pulse time was varied at a field strength of about $11·1$ kV/cm. A large number of visual and photographic measurements showed that with a 38 m cable the core had formed; with a 40 m cable sometimes half and sometimes complete tracks formed, and with a 41 m cable complete cathode-side tracks almost always formed. We calculated from this a mean rate of development of at least $1 \cdot 1 – 1 \cdot 3 \times 10^8$ cm/s, if the distance between the origin of the core and the cathode is taken as the distance to be traversed in accordance with the values given under section (b).

The anode-side track always runs exactly in the direction of the field lines, i.e. it takes the shortest path to the anode, but the half-developed cathode-side track often leaves the core at a slight angle to the field direction. In many cases it increases in breadth on its way to the cathode, and in the immediate neighbourhood of the cathode it can become a thin trace (*Figure 3(c)*). This point emits a white light when it is viewed in the dark.

In many cases where the track has extended across the whole region, 'branching' is seen to develop in the area of the original avalanche head, at a distance within 10% of the original head-to-cathode distance (*Figure 3(c)*).

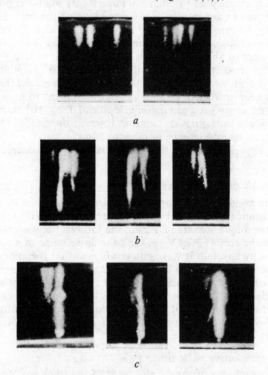

Figure 3. Development of the electron avalanche under the same conditions as in *Figure 2*, but the direction of the pulse has been prolonged until the normal avalanche has almost reached the anode. (*a*) avalanche head at the anode. If the voltage is raised, the cathode-side track closes up (*b*) to complete the side-to-side track (*c*).

3. *Points in Homogeneous Fields*

Examination of the photographs of the avalanche development leads to the impression that the tracks from the avalanche head to the anode or cathode behave like discharges travelling from a negative point to the anode (anode-side track) or from a positive point to the cathode (cathode-side track).

The following experiment was carried out to study the appearance of such discharges from a metal point in a homogeneous field, and if possible to measure the rate of growth of these discharges and compare them with the values given above.

A small metal sphere 1 mm in diameter and a roughly needle-shaped tip were used in a 3·6 cm wide homogeneous field. With a normal expansion ratio, in conditions of extensive scatter, as a rule only discharges surrounded by a cloud were obtained. The structure did not become visible until the expansion ratio had been reduced. Thus, using the same conditions as above (see section 2) the discharge from a roughly needle-shaped point projecting about 1 mm from the plate electrode was observed. Initiation of the tracks showed wide scatter (see *Figure 4*), and was lower for a negative point than for a positive one ($\approx$ 32 kV as against $\approx$ 37 kV). The onset of discharge was very irregular, and could not be improved by irradiation from a spark, so that the tracks were not reproducible and it was not possible to measure the velocity.

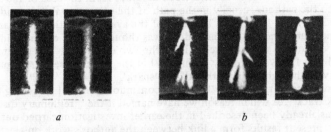

Figure 4. Discharges at a point in a homogeneous field (subnormal expansion ratio). Point length = 1 mm. In *a*: the point is below and negative. In *b*: point at top and positive. Voltage about 32 kV in *a*, 37 kV in *b*. Pulse duration 2·5 x 10^{-7} s.

However, the similarity of the discharges is nevertheless striking. The tracks from a *negative* point are non-branching and travel in straight lines as they do for the anode-side track starting from the avalanche head. The discharges from the *positive* point are branching, corresponding to the tracks in *Figure 3(b)*. Likewise, the positive discharge begins at higher voltages than the negative one, just as happened when the pulse time was kept constant and the cathode-side track appeared at a higher voltage than the anode-side. (Studies in an inhomogeneous field[4] have revealed the same properties and have shown also that the positive track develops at a higher rate (1 x 10^8 cm/s) than the negative (on average $\approx$ 2 x 10^7 cm/s).) This similarity means that the space charge in a homogeneous field produces an inhomogeneous field like the metal point.

Summary of results

The observations described here confirm the hypothesis based on observations of electron avalanches in their earliest stages, namely that the track of a normal electron avalanche is the same as that of the later spark. We may say that even with strong irradiation by ultraviolet light ($\approx$ 1 avalanche/cm^2 in 10^{-7} s; $\approx 10^{-12}$ A/cm^2) and 10 to 20% overvoltage, the breakdown of the gap is caused by a *single* avalanche without the help of neighbouring avalanches. The same processes that bring about the conversion of the normal electron avalanche to the 'spark' take place in this avalanche independent of the other avalanches. Thus the spark develops from the beginning in a narrow track, even in its very earliest stages, in contrast to the earlier theories according to which the development was over a wide front.

If we summarize the experimental results on the development of the avalanche in the spark track, we get the following picture: if we take a pulse time of some 10^{-7} s (electron distance a few centimetres) and do not exaggerate the overvoltage (10 to 20%), an electron avalanche will travel from the cathode into the gas space at about $1 \cdot 25 \times 10^7$ cm/s. If αx in the amplification factor exp (αx) has reached a value of about 20, the electron avalanche will be unstable and a track will start to develop from the avalanche head to the anode at a mean velocity of 7–9×10^7 cm/s (in my experiments: pressure ≈ 260 torr, overvoltage = 10 to 20%). If the anode-side kanal reaches the anode region, a track develops from the area of the avalanche head back to the cathode, usually in the path of the original avalanche and at a higher speed ($> 1 \cdot 2 \times 10^8$ cm/s). The head of the avalanche is thus the starting-point for the spark, and not the cathode as was thought. If the cathode-side track reaches the cathode, a narrow track bridges the two electrodes. This electron and ion channel is at first very slightly luminous, but the current rises very rapidly and the intensity of the light reaches that of the 'spark'.

The appearance of this weakly luminous continuous channel preceding the bright spark track (for which reason we have named it the 'preliminary discharge channel') has already been described in the earlier investigation carried out with Flegler. The present results form a link between the author's work on electron avalanches, i.e. the very earliest collision ionization processes, and other studies which have dealt with the progressive breakdown of the preliminary discharge track.

As already mentioned, the anode-side track is faintly luminous in a darkened room and that going from one side to the other is also luminous but brighter. These 'preliminary discharge tracks' seem to be identical with the 'hazy streaks' described by Dunnington:[5, 6] these are the first signs of the spark discharge seen during observations with an electro-optical shutter. In this way, cloud-chamber and electro-optical methods of investigation can supplement one another, since the latter can be used to study the formation of the bright spark track in the 'preliminary discharge track' (velocity of spark head and its direction of growth, start of bright spark within the track etc.) for which the cloud chamber is not suitable.

References

1. Raether, H., *Z. Phys.*, **107**, 91, (1937).
2. Kroemer, H., *Arch. Elekt.*, **28**, 703, (1934).
3. Flegler, E. and Raether, H., *Z, Phys.*, **103**, 316, (1936).
4. Flegler, E. and Raether, H., *Z. Phys.*, **11**, 435, (1935).
5. Dunnington, F. G., *Phys. Rev.*, **38**, 1535, (1931).
6. Hamos, L. V., *Ann. Phys.*, **7**, 857, (1930).

Introduction to Paper 4 (Raether, 'The development of kanal discharges')

This paper summarises the author's early formulation of a new theory of breakdown, which was largely a consequence of experimental observations of the type described in papers 2 and 3. The theory was developed independently of that proposed by Meek (given in paper 5). The two theories have much in common and have come to be known as the 'streamer theory' of spark breakdown. They are based on considerations of individual electron avalanches, the transition from an avalanche into a streamer and the subsequent behaviour of the streamer. The theory involves ionisation by electrons and photons in the gas and space-charge fields.

For air, on the assumption that the transition from an avalanche to a streamer occurs when the avalanche just crosses the anode/cathode gap, d cm, the breakdown criterion obtained is

$$\alpha d = 17 \cdot 7 + \log_e d$$

If there exists a critical avalanche length x_c at which the value of $\exp(\alpha x_c)$ is sufficiently high for the avalanche to develop into a streamer and if x_c is less than the anode/cathode separation, d, then the growth of the spark is expected to be by the streamer mechanism, whereas if $x_c > d$ the Townsend mechanism is expected to apply up to breakdown. According to Raether (*Electron Avalanches and Breakdown in Gases*, Butterworths, 1964) it can be shown that the processes described by the Townsend mechanism need to be completed by a space-charge effect for a large current growth and voltage drop to occur in the discharge gap.

The streamer theory has been extended and elaborated by a large number of authors whose contributions have been summarised in the references given below. See also papers 8, 9, 10 and 15.

Further Reading
Raether, H., *Electron Avalanche and Breakdown in Gases*, Butterworths, (1964).

Meek, J. M. and Craggs, J. D., *Electrical Breakdown of Gases*, Clarendon Press, (1953).

Loeb, L. B., *Basic Processes of Gaseous Electronics*, University of California Press, (1955).

Nasser, E., *Fundamentals of Gaseous Ionization and Plasma Electronics*, Wiley, 1971 (in particular, see chapter 9, sections 9.2-9.4).

4

The development of kanal discharges
H. RAETHER*

Introduction

The author's cloud-chamber observations have revealed the development of an electron avalanche in a homogeneous field into a spark 'kanal', and have shown how this process develops spatially and with time.[1] The following results have been obtained at ≈ 300 torr and an electrode separation of 3·6 cm: the normal avalanche discharge, which studies of the electron avalanche[2] have shown to travel at $1·25 \times 10^7$ cm/s in air ($E/p \approx 40$), reaches a critical amplification and develops further as a kanal discharge, with an increased velocity of $7-9 \times 10^7$ cm/s. After a kanal discharge reaches the anode it returns to the cathode with a velocity of $1-2 \times 10^8$ cm/s. Henceforth, a channel filled with charge carriers connects the anode and cathode. The luminous spark discharge occurs within this channel.

The purpose of the following remarks is to discuss the mechanism of this rapid formation of kanal discharges and, as far as possible, to explain it.

How the avalanche discharge becomes unstable and the anode-directed kanal develops

We will follow the development of an avalanche kanal in a homogeneous field E_0 up to high amplifications $\exp(\alpha x)$ (α = primary ionization coefficient, x = distance from the cathode) and shall investigate what differences from a normal electron avalanche occur when the field of the charges created is comparable with the original field E_0.

The avalanche head filled with electrons can be idealized to a negatively charged sphere, behind which is the positive space charge. The growth of the charge on the negative sphere during the development of the avalanche will continue at most until the opposing field between the negative avalanche head and the positive space charge left behind balances the original field (in fact this final value will be reached earlier, since collisional ionization will have come to an earlier end. However, we

* Reproduced from *Archiv für Electrotechnik*, **34**, 49 (1940) by permission of Springer-Verlag, Heidelberg.

will retain this idealization for the sake of simplicity). In this case the field at the part of the avalanche head facing the anode will amount to $2E_0$; higher values should not be possible. This observation of J. Sämmer[3] is a serious objection to all theories based on extensive distortions of the field in front of the avalanche head. It is rather surprising that no attempt has yet been made to discuss or contradict its validity when high field distortions are used in an explanation.

The following argument makes it possible to understand the formation of large field distortions in spite of Sämmer's objection. We first consider a negatively charged space-charge sphere (sphere radius = ρ) alone in the homogeneous field E_0. The field on the anode side of the sphere will be referred to as the 'front field', and that on the cathode side as the 'tail field'. Let the maximum front field be

$$E_0 + \Delta E_{max}$$

in which ΔE_{max} is less than E_0, so that we have the field pattern shown in *Figure 1*. Under the influence of this front field the 'front electrons' will ionize to a greater extent than in the undistorted field E_0. The question now is at what distance from the surface of the sphere does the tail field of the avalanche head of these front

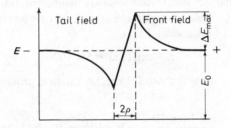

Figure 1. Shape of the field produced by superimposing a spherical field (ρ = radius of the sphere) on a homogeneous field, E_0 (section through the centre of the sphere).

electrons become comparable with the sphere field. If this happens at distances that are large in comparison with the sphere radius ρ, the front field has practically fallen back to E_0 and the field in front of the avalanche head does not rise above $2E_0$ in any of the succeeding electron avalanches (case A, *Figure 2*). However, if we get tail fields of the same magnitude as the sphere field at distances comparable with ρ, then the tail field will be greater than E_0, namely

$$E_0 + \Delta E_{max}(\rho/r)^2$$

(r = distance of the electrons from the centre of the sphere, $\Delta E_{max} \leqslant E_0$ [in *Figure 2* $\Delta E_{max} = E_0$], and the new front field ahead of the avalanche head of the front electrons can become

$$2E_0 + 2\Delta E_{max}(\rho/r)^2$$

(case B, *Figure 2*). If we imagine this schematized process as continuing, the front field can be a multiple of $2E_0$. *Figure 2* will make this process clear.

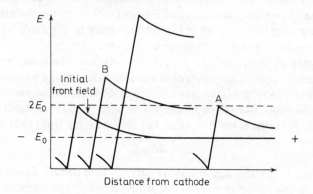

Figure 2. Diagram of growth of field above $2E_0$ ahead of the kanal tip.

During the cloud-chamber observations already mentioned, this 'becoming unstable' of the electron avalanche appeared at $\alpha x \approx 20$ (α = primary ionization coefficient, x = length of avalanche), i.e.

$$n = \exp(\alpha x) \approx 5 \times 10^8$$

The cross-section of the kanal is about 0·1 cm, i.e. the field produced by this space charge is

$$\Delta E_{max} = \frac{n \times 1 \cdot 6 \times 10^{-19}}{\rho^2} \times 9 \times 10^{11} \approx 30\ 000 \text{ V/cm}$$

whereas the undistorted homogeneous field amounts to 12 000 V/cm (*reference 1*, p. 468). Even though the numerical value of ΔE is uncertain, this estimate does show that ΔE and E_0 are of the same order of magnitude at the 'Kip' point of the avalanche.

Thus, talking very generally, a large sphere radius is favourable, since the 'range' of the sphere field which falls off in proportion to $(\rho/r)^2$ increases with ρ^2, the electrons therefore spend longer in the increased field. This case occurs during lightning discharges.

In what follows we shall take the case of a kanal discharge in a homogeneous field and attempt to work out the distance after which, in the field of a sphere, an electron avalanche attains a tail field comparable with this field. To apply these estimates to the cloud-chamber observations mentioned we adopt the following expression for α in the E/p range from about 40 to 180 (reference 4)

$$\frac{\alpha}{p} = A\left(\frac{E}{p} - B\right)^2 \qquad \begin{matrix} A = 1 \cdot 3 \times 10^{-4} \\ B = 28 \cdot 8 \end{matrix}$$

If n_0 electrons cross the surface of the sphere, there will be

$$n = n_0 \exp \left(\int_\rho^x \alpha dr \right)$$

electrons produced in the distance x. This electron avalanche travels in the field

$$E = E_0 + \Delta E_{max} \left(\frac{\rho}{r} \right)^2$$

and we can ask, as we did before, at what length x the charge of the avalanche head produces a field comparable with the sphere field. This leads to a relationship between x, the field strength at the 'kanal point' (surface of the sphere) and the number of initial electrons n_0. If the total field at the kanal tip is equal to $2E_0$, i.e. the initial field strength for the breakdown of the avalanche discharge into the rapid kanal discharge, we get the numerical values shown in *Table 1*.

Table 1

ρ	$n_0 = 1$	$n_0 = 10$	$n_0 = 1000$
	x	x	x
0·05	2	1·7	1·1
0·1	1·8	1·5	0·9
0·2	1·3	1·0	0·4
0·3	0·7	0·4	–
0·4	0·03	–	–

The calculation can be briefly outlined. Let the field strength at the head of the avalanche be

$$E_0 = \frac{n \times 1 \cdot 6 \times 10^{-19}}{4 \pi \Delta r^2}$$

$\left(n = \text{number of charges in the sphere } r, \Delta = \dfrac{1}{4 \pi \times 9 \times 10^{11}} \right).$

Using the above equations we get

$$\ln \left(\frac{n}{n_0} \right) = \ln \left(\frac{4 \pi \Delta r^2}{n_0 \times 1 \cdot 6 \times 10^{-19}} E_0 \right) = \int_\rho^x \alpha(E) dr$$

If now

$$E = E_0 + E_0 \left(\frac{\rho}{x} \right)^2$$

and assuming that $r = \rho$ (because the kanal cross-section is observed to be more or less constant), we can get a relationship between x and ρ. The above calculation was carried out with $E_0/p = 45$.

It was found that, when the kanal tip had a radius of 0·2 and 0·3 or above, a new tip developed at a distance $\approx \rho$ from the surface of the sphere, i.e. in a field which is higher than the undistorted field E_0. The number of initiatory electrons n_0 does not have a very great effect.

From observation, the radius of the kanal, and thus of its tip, is about 0·1 cm, so that remembering the assumptions we can speak of agreement here. The theory fits the observations even more closely when we recall that the calculation was carried out on the assumption that the 'front electrons' ionize by collision in a static sphere field. In fact, this sphere field travels behind the front electrons at a velocity approximately equal to the electron velocity.

To explain this process we can schematize it as follows: a sphere (avalanche head) filled with n_1^- electrons travels in a homogeneous electric field. The electrons ionize by collision after a certain distance (the ionization path) and n_1^+ positive and negative carriers are formed (*Figure 3*). The sphere thus contains $2n_1^-$ negative and n_1^+ positive charges, of which the $2n_1^-$ migrate towards the anode but the n_1^+ remains in position (*Figure 3(b)*). However from a certain charge density onwards (critical amplification), both the n^+ and n^- ions will tend to remain in place since the field between them approaches the original field (*Figure 3(c)*). In view of the small separation between the n^- and n^+ charges, the sphere field moves along with the migrating n^- electrons and thus increases the ionization yield. This accompanying migration of the sphere field was not allowed for in the calculation given above, and leads to lower values for the kanal tip radius ρ which are in still close agreement with observations.

Figure 3. Migration of the tip field with the front electrons. In (c), 1 attempts to retain 2, 2 therefore travels a little way behind 3, thus promoting collisional ionization by 3.

According to the picture, once the head of the avalanche has attained a certain amplification, it can give rise to a kanal discharge which at its head experiences field distortion greater than twice the value of the original field. (As the field distortion increases, so does the number of charge carriers per millimetre, so that the latter cannot be taken as constant in spite of 'space-charge deceleration'). At this stage of the theoretical argument it is hardly possible to provide data for the final value of the field distortion. The observed rates of growth in the cloud chamber are mean values. It is quite possible that measurements at greater electrode separations would show an increase in the rate of development, i.e. would reveal a further increase in the field distortion. Such an increase has been detected in the case of spark head velocities[5] and lightning predischarges.[6] However, we must bear in mind in this connection that we do not know in these cases how far the path was pre-ionized by a lower-current predischarge, which is either not luminous at all or too weakly so to be seen.

If we work on the basis of considerable field distortion at the tip of the kanal, the question arises whether the effect of this field distortion on the mobility of the electrons is sufficient by itself to account for the high rate of development, i.e. whether the kanal tip field can give the electrons their observed velocities. It has been shown by the author (reference 1, p. 480) that, taking into account the modification of the laws of mobility in high fields, velocities of 10^8 cm/s can be attained. On the other hand, the question arises whether photoelectrons produced by a gas-ionizing radiation[7] ahead of the kanal tip can help to accelerate the development.

To demonstrate that they do so we must show that an electron produced in the field of the avalanche head at a distance r greater than the sphere radius ρ will be amplified to such an extent by collisional ionization after travelling a distance Δr that a field similar to the original one at the tip is produced at the avalanche head of this photoelectron. The sphere field is thus advanced a distance $x = r + \Delta r$,

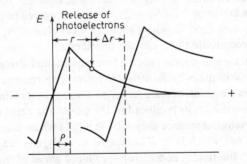

Figure 4. Explanation of the faster growth of kanal discharges when gas-ionizing radiation is active. Δr = path of a photoelectron released at a distance r in an electron avalanche, $r + \Delta r$ = forward displacement of the sphere field by the electron avalanche, the rate of development has thus increased to $\{(r + \Delta r)/\Delta r\}$ times the electron velocity.

although the electron has only moved by Δr. The ratio between the two distances gives the increase in the development velocity of the kanal as compared with the electron velocity (*Figure 4*).

The following calculation provides information concerning this problem. Let the undistorted homogeneous field be $E_0/p = 45$ (as above). Let the photoelectron be released at the distance $2\rho_1$. Then (the calculation proceeds as on p. 31) we find that for the field ahead of the kanal tip $\Delta E_{max} = \epsilon E_0$ at the distances x (see *Table 2*) from the tip of the original field.

Table 2

$\rho = 0.1$ cm

ϵ	1	3	4	8	10	12
x	$> 25\rho$	13ρ	6ρ	3.5ρ	2.7ρ	2.5ρ

These tabulated numbers indicate the following:

(a) when the field distortion is small (ϵ small) photoelectrons contribute nothing to the acceleration of the development of the kanal as the amplification of the photoelectrons to the point where a new kanal tip forms takes place well outside the original tip. The velocity increase factor is

$$\frac{r + \Delta r}{\Delta r} = \frac{x}{x - 2\rho} \approx 1$$

since $x > 2\rho_1$. We can conclude from this that the space charge alone is responsible for the sudden change of the normal avalanche discharge.

(b) On the other hand, when the field distortion is large, about ten times E_0, photoelectrons ahead of the kanal tip can give rise to an increase in kanal velocity as compared with the electron velocity. For example, when $\epsilon = 10$ ($\Delta E = 10E_0$) one gets an increase factor of $2.7/(2.7 - 2.0) \approx 4$ as compared to the electron velocity in this increased field. Given an adequate field distortion, therefore, photoelectrons do contribute to an increase in velocity.

These light quanta are of course emitted from the excited atoms in all directions. (The remarks made by R. Strigel in reference 8 concerning a preferred direction for this ionizing radiation are based on a misunderstanding.)

However, the photoelectrons produced to the side of the kanal tip contribute nothing to kanal development since they arise in a field which is only inappreciably distorted or even in one which is small compared with E_0, and the ionization yield falls off rapidly with the field. The electrons produced ahead of the kanal tip have the best yield provided they are not produced too far ahead of it. To determine the distance at which the photoelectrons are actually released we need not only the absorption coefficient μ of the ionizing radiation, but also the decrease in radiation intensity, and thus of ionization density, with distance. Since the ionization yield depends strongly on the field strength, only those light quanta which pass

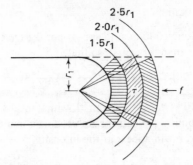

Figure 5. Calculation of the number of photoelectrons formed
by light quanta (from the surface of the kanal) which help to
promote the development of the discharge. These are effectively
found in the volume τ, since the ionization yield declines outside
the dotted lines.

through the surfaces f (within the dotted lines in *Figure 5*, and referred to as n_r)
and are absorbed within a distance dr produce photoelectrons of importance for
development. We get the following expression, therefore, for the ionization density ρ

$$\rho = \epsilon/\tau$$

(ϵ = number of electrons in the volume τ; as a first approximation the latter can, as a
function of r, be regarded as a constant).

$$\epsilon = \mu dr n_r$$

$$n_r = n_1 \left(\frac{r_1}{r}\right)^2 \exp\{-\mu(r - r_1)\}$$

(n_1 quanta pass through f at r_1) and thus with

$$\tau \cong \tfrac{4}{6}\pi r_1^2 dr$$

$$\rho(r) = \frac{\mu n_1}{\tfrac{2}{3}\pi} \left(\frac{1}{r}\right)^2 \exp\{-\mu(r - r_1)\}$$

or

$$\rho(r) \cong \rho_1(r_1)(r_1/r)^2 \exp\{-\mu(r - r_1)\}$$

i.e. the decrease in ρ is essentially determined by the absorption coefficient when

$$\mu > \frac{1\cdot4}{r_1} \quad (\text{at low } r, r = 2r_1)$$

[When r is very large however the absorption coefficient (p. 37) is the deciding
factor.] In air, $\mu \approx 1$ cm^{-1} (760 torr) or $\mu \approx \tfrac{1}{3}$ cm^{-1} (260 torr) so that this case
does not occur either with narrow kanal discharges, $r_1 \approx 0\cdot1$ cm, or with lightning
discharges, so that the quadratic term determines $\rho(r)$.

To give a numerical example:

We assume that only about 1000 quanta appear per 0·1 cm of path (i.e. several orders of magnitude less than ion pairs), so that

$$n_1 \approx 10^3(\tfrac{1}{6}) \quad (\pi r_1^2 \approx \tfrac{1}{6}4\pi r_1^2)$$

and thus $\rho(r_1) \cong 2·5 \times 10^3$ cm$^{-3} \approx 2·5$ mm^{-3}, which means that in the layer between r_1 and $1·5r_1$ about 3·5 electrons are produced in a tube $r = r_1$. Between $1·5r_1$ and $2r_1$ one electron will appear at the intensity assumed above, and $\mu = \tfrac{1}{3}$ cm^{-1}.

As already mentioned, the conditions for accelerated kanal development by photoelectrons are more favourable when the channel diameter is large, since the 'range' of the electric field (and thus the yield from collision ionization) increases with ρ^2 (ρ = radius of the kanal tip). As a result the influence of the photoelectrons increases more with the tip radius (ρ) than it does with increasing field distortion (ϵ). For example, under the conditions given above, and taking a tip radius of 1 cm for a tip field of $4E_0(\epsilon = 3)$, the extent of travel (Δr) of the electron must be 0·5 cm in order to reproduce the original field, whilst the kanal advances by $x = 1·5$ cm. This means the rate of development is already three times the electron velocity. We can conclude from this that the development of lightning tracks (towards the anode) involves photoelectrons playing a much greater part in kanal development than in that of the discharge tracks at small electrode separations.

Development of the kanal of the cathode side

The velocities measured during the reverse growth of the kanal on the cathode side were found to be so high that they could not be explained by the transit of positive ions; for explanation there only remain the photoelectrons produced by gas-ionizing radiation (see reference 1, p. 483). These photoelectrons thus have the initial task of producing the ionization track to the cathode; moreover an estimate must be made of how far they can affect the velocity. Both these points are discussed in the following.

In contrast to the development of the kanal on the anode side, there is already in this case a considerable distortion of the field due to the positive space charge formed by the anode-directed kanal at the end facing the cathode. The electrons migrating in this section of the kanal give rise to short-wave quanta which amongst other things release photoelectrons in front of this positive space-charge tip and thus initiate the cathode-directed kanal discharge.

This explains the experimental finding that the cathode-directed kanal only begins to develop after the onset of the anode-directed kanal. (It is conceivable that under certain circumstances the cathode-directed kanal can begin before the anode-directed kanal reaches the anode. It will only be possible to comment on this after experiments at greater electrode separations). If we assume the effect of the photoelectrons to be as described above, and take it that an electron released at a distance 2ρ from the tip of the kanal (measured from its centre), will move in the reverse direction, i.e. into an increasing field, then at $\epsilon = 10$, i.e. the field at the kanal tip (radius 0·1 cm) is equal to ten times the strength of the undistorted field, we obtain a path of $0·3\rho$ before there is enough distortion of the field ahead of the avalanche tip. The tip of the channel is thus advanced by $0·7\rho$. The rate of development is thus about 2–3 times greater than the electron velocity. If we bear in mind that the photoelectron moves against the positive space charge and thus into a rising field, the field distortion will not only be retained but will actually increase. The branching observed in the cathode-side kanal (reference 1, p. 473) together with the higher rate of development can also be ascribed to a higher tip field strength.

Development of lightning discharges

The above arguments make it possible to explain the kanal development clearly observed in the cloud chamber as a primary ionization process. Schonland *et al.*[6] have made observations on lightning discharges in which the 'principal discharge' was found to have a velocity of more than 10^9 cm/s. Strigel[8] has found similar high values in the 'principal discharge' between a point and plane. It is difficult to interpret such velocities in the manner given above. We must rather take into account the radial field of the discharge channel which is formed by the 'preliminary discharge'. As was shown in reference 1, p. 486, such a radial field will act on the motion of the electrons like a longtitudinal field, in that during the free path lengths it drives the laterally scattered electrons into the longtitudinal field direction. We will now assume that collisional ionization takes place at an arbitrary point in the channel (see *Figure 6*). This gives rise to ionizing radiation and thus to photo-electrons to an extent corresponding to the absorption coefficient. In the case already dealt with (p. 34) these photoelectrons created extensive distortion of the tip field by collisional ionization, so that the longitudinal field was steadily advanced. In this present case, the transverse field which increases the electron mobility in the same way, is already present owing to the 'preliminary discharge'. Thus even the more remote photoelectrons find themselves in a transverse field in which they can soon radiate short-wave quanta accompanied by an increased collision ionization yield. It is difficult to give anything other than very indefinite figures for this increase in velocity, but it can be readily seen that the high velocities of over 10^9 cm/s for the principal discharge can be readily explained in this way. *Figure 6* shows the general case of simultaneous forwards and backwards development, any point on the discharge track being taken as the point of origin.

The 'cathode-directed' kanal of the spark finds no carrier channel available in this sense, since the number of ions given by the previous avalanche discharge is too small to build up a transverse field and practically all the electrons are found in the avalanche head. Thus both the anode and the cathode-directed kanals are primary ionization processes independent of the previous history. The same will apply to what is called the 'preliminary discharge' of lightning (in contrast to the much brighter 'principal discharge') for which reason their rates of development are comparable.

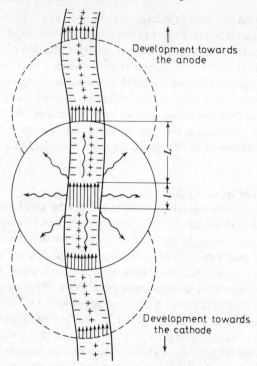

Figure 6. Illustration of the rapid forward development of a discharge in a carrier tube formed by a preliminary discharge
l = path of electrons from which sufficiently short wave quanta will be emitted to release photoelectrons in the distance L. L/l then gives the increase in velocity of the discharge as compared with the electron mobility.

Conclusion

The behaviour of an electron avalanche in a homogeneous field on its passage to the anode is discussed. The number of charge carriers increases exponentially ($n = \exp(\alpha x)$) and the velocity is the normal avalanche velocity $u = 1\cdot25 \times 10^7$ cm/s (air, $E/p \approx 40$). When a certain critical amplification is reached an increasing distortion of the field starts to build up in front of the avalanche head ('growing instability' of the avalanche discharge and transition to an 'anode-directed kanal'). As the field distortion increases the photoelectrons produced by the gas-ionizing radiation ahead of the kanal tip also help to increase the rate of development. If high field distortion is present, then with the aid of photoelectrons the discharge will also return to the cathode (cathode-directed kanal). On the conclusion of this counter-growth from the avalanche head to the cathode, the two electrodes are joined by a conducting tube, and this represents the subsequent 'spark channel'. A very rapidly developing discharge can form in such a conducting channel as a consequence of the transverse field and with the aid of the gas-ionizing radiation.

References
1. Raether, H., *Z. Phys.* **112**, 464, (1939).
2. Raether, H., *Z. Phys.* **107**, 91, (1937).
3. Sämmer, J., *Z. Phys.* **83**, 814, (1933).
4. Jodlbauer, A., *Z. Phys.* **92**, 116, (1934).
5. Holzer, W., *Z. Phys.* **77**, 680, (1932).
6. Schonland, B. F. J., Malan, D. M., and Collens, H., *Proc. Roy. Soc.* **A152**, 617, (1935).
7. For the evidence of this radiation, see Raether, H., *Z. Phys.* **110**, 611, (1938).
8. Strigel, R., *Wiss. Veröff, Siemens-Werk.* **15**, 13, (1936).

Introduction to Paper 5 (Meek, 'A theory of spark discharge')

The discharge theory given in this paper was developed by Meek independently of that given in paper 4 by Raether. The paper followed earlier ones by Loeb and Leigh, and by Loeb and Kip, in which a qualitative picture of breakdown involving the formation of 'streamers' was given. Loeb's early work was stimulated by his observations and those of Kip of the streamers produced in corona discharges (see paper 19).

As will be seen from the final section of Meek's paper, it was claimed that the proposed 'streamer theory' was capable of accounting for the experimental data then available on the breakdown process at high pressures or large gap distances in uniform electric fields. The need for an alternative theory to the Townsend theory was later disputed by Llewellyn Jones and his co-workers (see paper 6).

Further Reading

Loeb, L. B. and Leigh, W., *Phys. Rev.* **51**, 149A, (1937).

Loeb, L. B. and Kip, A. F., *J. Appl. Phys.* **10**, 142, (1939).

Loeb, L. B. and Meek, J. M., *Mechanism of the Electric Spark*, California: Stanford University Press, (1941).

Nasser, E., *Fundamentals of Gaseous Ionization and Plasma Electronics*, Chapter 9, Sections 9.2–9.4, Wiley (1971).

5

A theory of spark discharge

J. M. MEEK*

Introduction

In the early 1900s J. S. Townsend[1] evolved his now classical theory of the mechanism of spark discharge on the basis of measurements made at high values of E/p, the ratio of field strength to pressure, and low values of pd, the product of pressure times gap-length. His theory has been extrapolated to explain the mechanism of sparking under ordinary conditions for longer gap-lengths at atmospheric pressure and above. It has formed the guiding principle of all considerations of the sparking mechanism for nearly 40 years. Many researches have indicated that for values of pd near and somewhat above the minimum sparking potential the theory was adequate and relatively accurate. The theory in this region enables calculation of the sparking potentials to be made with adequate accuracy, and recently it has been shown by Schade[2] to be consistent with time-lag studies.

In recent years, however, it has become increasingly apparent that the theory is entirely inapplicable when extrapolated to sparks for pd greater than about 200 mm Hg x cm in air, and analogous values in other gases. The reasons for this conclusion may be stated briefly as follows:

1. The formative time lags of sparks at *atmospheric pressure*, i.e., the time required for the breakdown mechanism to materialize, have been measured by several diverse methods, and all lead to time intervals of the order of 10^{-7} s, or even less, for a 1 cm gap at small overvoltages.[2] These time intervals are far below the microseconds required by positive ion movement as shown by Schade[3] and others.[4]

2. At atmospheric pressure the sparking potential has been found to be independent of cathode material to a very high degree,[5] while the theory above, even though somewhat insensitive to the value of γ at higher pressures, requires a definite dependence of the order of 10 per cent on the value of γ.[6] The values of pd at which this dependence begins rapidly to vanish occur again at $pd \approx 200$ (reference 5).

* Reproduced from *The Physical Review,* 57, 722, (1940) by permission of the American Physical Society, New York.

3. In longer sparks, in lightning discharge, and in positive point corona at atmospheric pressure, it is virtually certain that the character and action of the cathode is without any influence whatsoever on the process.

4. The observed appearance of sparks in measurements of Townsend's coefficient, α, at higher pressures and adequate x with no measurable value of γ indicates in another way that the theory based on a γ is no longer applicable at higher values of pressure and gap length.[7]

5. Observations of actual sparks in short time intervals made visible by Kerr cell-shutter studies[8] and by cloud-track pictures of spark paths after short time intervals,[9] as well as the usual visual observations of sparks, indicates that breakdown at higher pressures occurs along very narrow channels or filaments. This is incompatible with the transient rearrangement of conduction currents in a gas caused by accumulations and movements of positive ions or successive waves of electron ionization when $\gamma \exp (\alpha d) > 1$. This discrepancy is made still more glaring by the zigzag nature of the spark filament and the frequent branching observed in longer sparks.

6. While only one electron is required to initiate a spark on the Townsend theory as interpreted by Holst and Oosterhuis[10] it is clear that the greater the number of photons incident on the cathode by external irradiation the more rapidly should the space charge accumulations occur, so that formative time lags should on that theory decrease as the illumination intensity of the cathode increases. This has been shown to be true by Schade for low values of *pd*. However, at high *pd* observation shows no appreciable change in the *formative* time lag and of the sparking potential with illumination over a considerable range of intensity.[11] When the illumination intensity is increased by a factor of 10^5 beyond that giving the 10^7 electrons/s/cm^2 generally used, both formative time lag and sparking potential are lowered.[12] In this case the potential is lowered at a maximum by no more than 10 per cent, and the formative time lag appears to be doubled when the intensity of illumination is decreased by a factor of 500.

Theory

The above-mentioned objections to the classical theory make it imperative to formulate a new mechanism which does not so conflict with experimental observations. The importance of streamer formation in the mechanism of spark breakdown has been discussed in a recent article by Loeb and Kip[13,13a] and more extensively by Loeb in his book *Fundamental Processes of Electrical Discharge in Gases.* The breakdown process is there pictured in qualitative fashion, but no quantitative criterion is given for streamer formation and breakdown. However, it is shown that under sparking conditions the positive space charge in an electron avalanche which is originated at the cathode produces an electric field of the same order as the external applied field when the avalanche reaches the anode. This fact was also indicated previously by Slepian,[14] who then postulated the development of thermal ionization in the avalanche. It is further stated by Loeb that the axial field distortion produced by the positive space charge inhibits the advance of the electron avalanche towards the anode, but favours the development of a positive streamer towards the cathode. An explanation of the mechanism of streamer formation and mid-gap streamers is also given, though the fundamental criterion for their appearance is lacking. However, it is indicated that breakdown will not occur until a conducting filament exists across the gap. The single electron avalanches observed by Raether[15] just below breakdown do not form such a filament but produce a charge concentration at the anode so that positive streamer propagation to the cathode is needed.

In the above-mentioned theoretical discussions, consideration appears to have been given principally to the field distortion produced by the positive space charge in the direction of the axis of the avalanche, i.e., in the direction of the external applied field. Clearly, however, the positive space charge will also produce a field distortion in the direction radial to the axis of the avalanche, i.e., perpendicular to the external field. Consideration of this fact led the writer to the following criterion for the transition of an electron avalanche to a conducting streamer, a criterion which makes quantitative calculation possible. *A streamer will develop when the radial field about the positive space charge in an electron avalanche attains a value of the order of the external applied field.* Then photoelectrons produced in the gas in the immediate vicinity of the avalanche will not only be accelerated in the direction of the external applied field but will also be drawn into the stem of the avalanche. (It has been shown by Cravath,[16] Dechene,[17] Loeb and Kip[13] that streamer propagation depends on photo-ionization in the gas.) In this manner the positive ion space charge attracts photoelectrons from the surrounding space in sufficient numbers to originate a self-propagating streamer. For lower values of external applied field, where the multiplication of ions in the avalanche is insufficient to produce a radial field of the same order of magnitude as the external field, photoelectrons produced in the gas proceed to the anode and are not deflected to the main avalanche.

It is clear that the exact equality of the radial field and the external applied field need not be insisted upon, and it is possible that the requisite criterion could be satisfied when the radial field is only 50 per cent of the applied field. However, in the calculations the two fields have been set equal, and the application of this condition at once leads to equations for field strength for spark breakdown that closely approximate experimental observations and clarifies difficulties in the qualitative picture given by Loeb. The criterion not only enables quantitative calculations of breakdown to be made, but it also assists in the interpretation of a number of other sparking phenomena. It thus seems of interest to present the results of the analysis and its applications, together with its experimental justification.

Derivation of equation for breakdown
Because of the cumulative character of the ionization in an electron avalanche and the relative immobility of positive ions as compared with electrons, the bulk of the positive ions left behind by the advancing electron swarm are concentrated in a region of several ionizing free paths behind the tip of the avalanche. The region

is also bounded by the periphery of the avalanche, which is conditioned by electron diffusion. Since the estimation of the field produced by such a distribution is difficult it will be assumed for convenience of calculation that the positive space charge is concentrated in a spherical volume of radius, r, equal to that of the avalanche. While the volume may be more nearly in the form of an oblate or prolate spheroid, the assumption of a perfect sphere will not affect the calculated field in order of magnitude.

The field, E_1, at the surface of a spherical volume of uniformly distributed space charge is given by

$$E_1 = \frac{Q}{r^2} = \frac{4}{3}\pi r N \epsilon,$$

where N is the number of ions per cubic centimetre and ϵ is the charge on an electron. To find N, one recalls that owing to cumulative ionization the number, n, of positive ions formed when the avalanche has progressed a distance, x, is $n = \exp(\alpha x)$, where α is the Townsend coefficient for ionization by electrons. In a distance dx the number of ions created is $dn = \alpha \exp(\alpha x)\, dx$. These ions are assumed to be contained in a cylindrical volume of radius, r, and of length, dx, so that

$$N = \frac{\alpha \exp(\alpha x)}{\pi r^2}.$$

According to Raether[15] the radius, r, is related to the distance, x, of avalanche travel by means of the diffusion equation $r^2 = 2Dt = 2Dx/v$, where v is the velocity of advance of the avalanche and has been measured by Raether[15] and White.[18] Since $v = KE$, where K is the mobility and E is the applied field, we have

$$E_1 = \frac{4}{3}\frac{\alpha \exp(\alpha x)\epsilon}{r} = \frac{4}{3}\frac{\alpha \exp(\alpha x)\epsilon}{(2Dx/KE)^{\frac{1}{2}}}.$$

Both D and K are functions of E/p and are approximately related by the expression

$$\frac{D}{K} = \frac{P}{N\epsilon}\frac{c_1^{\,2}}{c^2},$$

where P is atmospheric pressure, $N\epsilon$ is the Faraday constant, c is the speed of thermal agitation of electrons ($1\cdot2 \times 10^7$ cm/s at $20°C$), and c_1 is the square root of the mean-squared velocity of the electron in the applied field. Townsend[19] has shown that c_1 is considerably larger than c. According to Compton[20] the value of c_1 is given by

$$c_1^{\,2} = 1\cdot33\frac{\epsilon}{m}\frac{E\lambda}{\sqrt{f}},$$

where λ is the mean free path of the electron, E is the applied electric field, m is the mass of an electron, p is the gas pressure, and f is a factor which depends on the fractional loss of energy per collision. If λ_0 is the mean free path of an electron at a pressure of 760 mm Hg, then the value of c_1 for a pressure, p, is given by

$$c_1{}^2 = 1010 \, \frac{\epsilon}{m} \frac{E}{p} \frac{\lambda_0}{\sqrt{f}}.$$

The values of λ_0 and f obviously depend on E/p and are not known for the high values of E/p which are necessary for spark breakdown. But inasmuch as the theory will be found insensitive to slight errors in the value of $\lambda_0/\sqrt{f}$ it is considered reasonable to calculate its value from known data based on experiments with fields close to those which produce sparking. Raether[15] found that the speed of advance of an electron avalanche was $1 \cdot 25 \times 10^7$ cm/s in a field $E/p = 41$ V/cm/mm. Substitution of such values in the well-known Langevin equation for electron mobility, *viz.*, $v = 0 \cdot 815 \, \epsilon \lambda E/mc_1$, enables one to calculate that $\lambda_0 \sqrt{f} = 5 \cdot 7 \times 10^{-6}$. The mean free path λ_0 is taken as $3 \cdot 6 \times 10^{-5}$ from curves given by Brose and Saayman[21] and corresponds to an electron energy of three volts. Whence we have $f = 0 \cdot 025$ and $c_1 = 2 \cdot 01 \times 10^7 (E/p)^{\frac{1}{2}}$. Then we have

$$E_1 = \frac{4}{3} \frac{\alpha \exp(\alpha x) \, \epsilon}{(2Pc_1{}^2 x/N\epsilon c^2 E)^{\frac{1}{2}}}$$

$$= 5 \cdot 28 \times 10^{-7} \frac{\alpha \exp(\alpha x)}{(x/p)^{\frac{1}{2}}} \text{ volts/cm.} \tag{1}$$

Now for $x = d$, the gap length for a plane parallel gap, application of the criterion for streamer formation, that $E_1 = E$, gives

$$E(d/p)^{\frac{1}{2}} = 5 \cdot 28 \times 10^{-7} \alpha \exp(\alpha d)$$

which may be written in logarithmic form as

$$\frac{\alpha}{p} pd + \log_e \frac{\alpha}{p} = 14 \cdot 46 + \log_e \frac{E}{p} - \tfrac{1}{2} \log_e pd + \log_e d. \tag{2}$$

Reference to curves of the Townsend type which relate α/p with E/p now enables the value of E/p required for sparking to be calculated for given d and p. In the case of a 1 cm gap in air at atmospheric pressure the breakdown equation is satisfied for $E/p = 42 \cdot 4$ ($\alpha = 18 \cdot 6$), i.e., $E = 32 \cdot 2$ kilovolts per cm,* which is in close agreement with the observed value of $31 \cdot 5$ kilovolts per cm.[22]

Comparison between calculated and measured values of breakdown potential
The form of the breakdown eqn. (2) shows that it is in agreement with Paschen's law other than for the term, $\log_e d$. This term has little effect, however, on the calculations, as shown by *Table 1* for breakdown potential corresponding to different values of p and d. The deviations, for the same value of pd, are within the present margins of experimental error in measurement, and it may be that for

---

* If we set the criterion for streamer formation as $X_1 = 0 \cdot 2X$ instead of $X_1 = X$, the breakdown field strength evaluted is $31 \cdot 8$ kilovolts per cm.

Table 1—*Breakdown potentials corresponding to different values of p and d.*

pd mm × cm	Breakdown potential (kV)		
	$d = 0{\cdot}1$	$d = 1{\cdot}0$	$d = 10{\cdot}0$
7600	245	248	249
760	31·5	32·2	32·9
380	18·0	18·6	18·9

more accurate measurements such differences will be observed. For it should be pointed out that the derivation of Paschen's law is based on theoretical considerations analogous to those of Townsend. Such considerations make the mechanism of spark breakdown dependent on the total number of ions formed in the gap and, since the multiplicative factors in ion production depend on E/p, Paschen's law results. However, on the new theory, it is seen that the breakdown mechanism is not dependent on the total number of ions formed but depends on the ion density in an electron avalanche and the resultant space-charge field. In consequence, we would not expect that Paschen's law would be strictly obeyed. In fact, it has previously been shown[23] that with space charge distortion Paschen's law is only approximately correct, and that the deviation in breakdown potential is of the same order of magnitude as that derived here. So that unless the precision of experimental determinations of sparking potentials in the future indicates Paschen's law to hold beyond the prediction of the present theory, the deviations of the latter from Paschen's law are not thought to invalidate the theory in any way.

Both experimental and calculated curves to show the relationship between the voltage required to cause breakdown and the product, *pd*, of pressure times gaplength are plotted on logarithmic scale in *Figure 1*. The calculated curve is based on values for the primary ionization coefficient, α, in air as given by Sanders[24] for E/p up to 100; for higher E/p the values of α are taken from Posin's observations

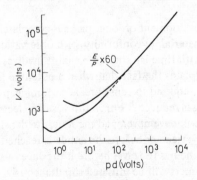

Figure 1. Experimental and calculated curves to show relationship between breakdown voltage and *pd*.

in nitrogen.[25] The experimental curve, shown dotted, is that given by Whithead[26] as the mean of the results given by a number of independent workers. It will be observed that the calculated curve is in agreement with that experimentally determined for values of pd from 10 000 down to about 100 mm x cm. For values of $pd < 100$, the calculated values are higher than those observed, and the deviation increases steadily with decreasing pd and thus with increasing E/p. The deviation occurs in the region of $E/p = 60$ which is about the minimum value of E/p for which the second Townsend coefficient is observed.[27] Thus for values of $pd < 100$ the deviation can be accounted for by the presence of other sources of ionization in the gap. This is in agreement with calculations made in the past on a knowledge of the Townsend coefficients, and which have yielded approximately correct values of breakdown potential corresponding to low values of pd. The deviation between the two curves then indicates the region in which breakdown by streamer formation gives way to classical theory and the Townsend type of mechanism.

Preliminary calculations of the breakdown strength of other gases, such as hydrogen, have been made and give values in agreement with those observed experimentally.

Conclusions

It is seen that the above theory postulates an entirely different mechanism of breakdown for longer gaps and higher pressures than that given by classical theory.

1. The proposed mechanism depends on photo-ionization and electron multiplication in the gas (Townsend's α). This photo-ionization is justified by recent observations on the effectiveness of such ionization under the existing conditions.[16, 17]

2. It requires that the density of positive ion space charge created in an electron avalanche is sufficient to draw in photoelectrons and their progeny, produced by ionization by collision, in a measure which renders the positive space charge a conducting plasma, and permits the further development of the streamer. This is only possible when the field strength and gas pressure are such as to allow a sufficient density of photo-ionization in the neighbourhood of the streamer head to permit self-propagation.

3. It shows that the development of long sparks is completely devoid of any dependence on cathode material, in conformity with observation.

4. It is in full accord with time-lag studies in that it involves formative time lags of $\approx 10^{-7}$ s or less and explains the fact that when a potential is applied to a gap in excess of the minimum required to cause breakdown midgap streamers develop and the formative time lag is reduced.[8] For in this case an electron avalanche which originates at the cathode will have developed the requisite density of positive ions to cause the initiation of a streamer before the anode is reached. The reduced distance which the avalanche has to travel before it produces a streamer makes it probable that several streamers will be initiated simultaneously. A conducting filament across the gap will be formed by the junction of such streamers, and results in a corresponding reduction in time of formation of the spark.

5. At the sparking threshold the spark is initiated by a single electron liberated from the cathode. The chance of a spark then depends on (*a*) the chance of a single electron avalanche which produces at the anode or in midgap the requisite space charge, and (*b*) the success which this space charge has in photo-ionizing the ambient gas and in attracting sufficient electrons to ensure self-propagation of a streamer. This accounts for the statistical time lags, and for the observed indefinite character of the sparking potential. That the sparking threshold is at all definite under these conditions is explained by the cumulative nature of electron ionization, i.e., $i = i \exp (\alpha x)$, and also that $\alpha/p = f(E/p)$ where $f(E/p)$ is of the form $A \exp (BE/p)$ or $C(E/p - D)^2$.

6. The theory is completely consistent with the observations of H. Raether on the development of sparks in a cloud chamber. In fact, Raether's photographs clearly indicate the passage of an electron avalanche from cathode to anode followed by a streamer which proceeds from the anode to the cathode to form a conducting filament bridging the gap. Simultaneously and quite independently, Loeb, on the basis of corona studies, and Raether,[28] on the basis of cloud-chamber observations, postulated a qualitative picture of the sparking mechanism which is here rendered quantitative.

7. It can be shown that the field distortion produced by cumulative ionization with high densities of photoelectric current gives a lowering of sparking potential in rough quantitative agreement with the lowering observed by White,[11] Rogowski and Wallraff.[29]

8. It clearly accounts for the filamentary character of spark breakdown at longer gaps and higher pressures and explains both the zigzag character of the spark and the branching observed. This differentiates the sparks under this mechanism from those in the Townsend regime ($pd < 100$ in air).

9. It explains the fact that in the study of the Townsend coefficients at lower E/p and higher pd sparking occurs before values of the second coefficient, γ, can be observed.

10. The slight deviation of the theory from Paschen's law is not inconsistent with experimental data in this field of investigation.

11. While at present the inexact character of the quantitative criterion for streamer formation leaves considerable latitude owing to the exponential nature of electron multiplication by ionization by collision, the equation gives numerical results in as good agreement with experiment as those based on Townsend's theory, where a hypothetical value of γ has to be assumed, since it cannot be observed. Numerical agreement of this character between theory and experiment must not be taken too seriously as a confirmation of any particular theory in view of the latitude given by the exponential term and the lack of adequate experimental data.

12. The theory has been extended to the explanation of sparks of all lengths at higher pd in air and the necessary conditions for sparking have been established by Loeb. A more exhaustive discussion of the application of the theory will be given in an article to appear at some future date.

13. It can be shown that the theory leads to an explanation of the breakdown of

unsymmetrical gaps where the field distribution is known, and that quantitative calculations can be made. The theory may be applied to either surge-impulse or static breakdown.

14. It is found that the self-propagating streamer mechanism, as modified for longer sparks, will adequately account for the pilot streamer in Schonland's mechanism[30] for the lightning discharge, and it clarifies the theory proposed by the writer[31] for the 'stepping' observed in lightning and spark discharges.

The writer wishes to express his appreciation to Professor L. B. Loeb, whose original qualitative theory stimulated this work and whose suggestions and criticisms have contributed much to the development of the present theory. He also thanks the Commonwealth Fund for the grant of a Fellowship, during the tenure of which the work was carried out.

References and Notes

1. Townsend, J. S., *Nature* 62, 340, (1900); *Phil. Mag.* 1, 198, (1901).
2. Pedersen, P. O., *Ann. d. Physik* 71, 371, (1923); Beams, J. W., *J. Frank. Inst.* 206, 809, (1928); Torok, J. J., *Trans. A.I.E.E.* 47, 177, (1928); 48, 46, (1930); Buraway, *Arch. f. Elek.* 16, 14, (1926); Tamm, I., *Arch. f. Elek.* 19, 235, (1928); Rogowski, W., *Arch. f. Elek.* 20, 99, (1928); Dunnington, F. G., *Phys. Rev.* 38, 1535, (1931); White, H. J., *Phys. Rev.* 46, 99, (1934); Tilles, A., *Phys. Rev.* 46, 1015, (1934); Wilson, R. R., *Phys. Rev.* 49, 1082, (1936); Newman, M., *Phys. Rev.* 52, 652, (1937); Strigel, R., Wiss. Veröffentl. Siemens Werken 15, 1, (1936).
3. Schade, R., *Zeits. f. Physik* 104, 487, (1937).
4. Tamm, I., *Arch. f. Elek.* 19, 235, (1928); Rogowski, W., *Arch. f. Elek.* 20, 99, (1928); Rogowski, W. and Tamm, I., *Arch. f. Elek*, 20, 625, (1928); Tilles, A., *Phys. Rev.* 46, 1015, (1934).
5. Loeb, L. B., *Fundamental Processes of Electrical Discharge in Gases* (John Wiley, 1939), p. 415.
6. Reference 5, p. 416 *et seq.*
7. Sanders, F. H., *Phys. Rev.* 44, 1020, (1933); Posin, D. Q., *Phys. Rev.* 50, 650, (1936); Bowls, W. E., *Phys. Rev.* 53, 293, (1938).
8. von Hamos, L., *Ann. d. Physik* 7, 857, (1930); Dunnington, F. G., *Phys. Rev.* 38, 1535, (1931); White, H. J., *Phys. Rev.* 46, 99, (1934).
9. Flegler, E., and Raether, H., *Zeits. f. Physik* 99, 635, (1936); 103, 315, (1936); Raether, H., *Zeits. f. Physik* 96, 567, (1936); 110, 611, (1938), 112, 464, (1939).
10. Holst, G. and Oosterhuis, E., *Phil. Mag.* 46, 1117, (1923).
11. White, H. J., *Phys. Rev.* 48, 113, (1935); Wilson, R. R., *Phys. Rev.* 50, 1082, (1936).
12. White, H. J., *Phys. Rev.* 48, 113, (1935).
13. Loeb, L. B. and Kip, A. F., *J. App. Phys.* 10, 142, (1939).
13a. At the request of Professor Loeb, attention is called to page 157 of this article where a careless blunder is made in the calculation of the field strength

produced by space charge. However, the order of magnitude of the calculated values is not affected by correction, and the argument given in the article is by no means invalidated thereby.

14. Slepian, J., *Electrical World* **91**, 761, (1928).
15. Raether, H., *Zeits. f. Physik* **107**, 91, (1937).
16. Cravath, A. M., *Phys. Rev.* **47**, 254, (1935).
17. Dechene, C., *J. de phys. et rad.* **7**, 533, (1935).
18. White, H. J., *Phys. Rev.* **46**, 99, (1934).
19. Townsend, J. S., *Electricity in Gases* (Oxford University Press, 1914), p. 122 ff.
20. Compton, K. T. and Langmuir, I., *Rev. Mod. Phys.* **2**, 220, (1930).
21. Brose, H. L. and Saayman, E. H., *Ann. d. Physik* **5**, 797, (1930).
22. Spath, W., *Arch. f. Elek.* **12**, 331, (1923).
23. Varney, R. N., White, H. J., Loeb, L. B. and Posin, D. Q., *Phys. Rev.* **48**, 818, (1935).
24. Sanders, F. H., *Phys. Rev.* **44**, 1020, (1933).
25. Posin, D. Q., *Phys. Rev.* **50**, 650, (1936).
26. Whitehead, S., *Dielectric Phenomena* (D. Van Nostrand & Company, New York, 1927), p. 42.
27. Bowls, W. E., *Phys. Rev.* **53**, 293, (1936).
28. Raether, H., *Zeits. f. Physik* **112**, 464, (1939).
29. Rogowski, W. and Wallraff, H., *Zeits. f. Physik* **102**, 183, (1937).
30. Schonland, B. F. J., *Proc. Roy. Soc.* **A164**, 132, (1938).
31. Meek, J. M., *Phys. Rev.* **55**, 972, (1939).

Introduction to Paper 6 (Llewellyn Jones and Parker, 'Electrical breakdown of gases. Part 1: Spark mechanism in air')

This paper was the first of a series of papers by Llewellyn Jones and his associates which attempted to show that for uniform-field conditions at high values of pd (p = gas pressure, d = anode/cathode separation) the spatial growth of the pre-breakdown current could be represented, as at low values of pd ($\lesssim 150$ torr cm), by the expression

$$I = I_0 \exp{(\alpha d)} / [1 - (\omega/\alpha)\{\exp{(\alpha d)} - 1\}]$$

The paper makes no allowance for the occurrence of electron attachment, (see papers 11 and 12) but this does not invalidate its major conclusions. Perhaps the major contributions of this paper are:

(i) the careful analysis it makes of the experimental conditions which need to be satisfied if the growth of pre-breakdown currents is to be examined in a meaningful way,

(ii) the demonstration that, if the experimental investigations are done with sufficient care, 'upcurving' of the log I/d curves can be observed for conditions near breakdown. This implies, as is explained in the paper, that a secondary ionisation mechanism operates throughout the breakdown process and that no new mechanism needs to be introduced just before breakdown occurs.

More recently, Dutton and Morris (*Brit. J. Appl. Phys.* **18**, 1115, (1967)) have extended the earlier work to breakdown voltages for air of $\gtrsim 400$ kV at values of pd up to 12 600 torr cm. Similar work for nitrogen was reported at the Ninth International Conference on Phenomena in Ionised Gases, Bucharest (1969) by Dutton *et al.* Boyd, Bruce and Tedford (*Nature*, **210**, 719, (1966)) have also reported similar measurements. It was found that even at these high values of pd the generalised Townsend equation was capable of correctly describing the observed spatial growth of current.

Further Reading

Dutton, J., Haydon, S. C., and Llewellyn Jones, F., *Proc. Roy. Soc.* **A213**, 203, (1952).

Llewellyn Jones, F., *Ionization and Breakdown in Gases*, Methuen, (1966).

Dutton, J. and Morris, W. T., *Brit. J. Appl. Phys.* **18**, 1115, (1967).

Boyd, H. A., Bruce, F. M., and Tedford, D. J., *Nature* **210**, 719, (1966).

6

Electrical breakdown of gases.
Part 1: Spark mechanism in air

F. LLEWELLYN JONES AND A. B. PARKER*

Introduction

There is at present no satisfactory quantitative explanation of the electrical breakdown of gases at values of the parameter pd greater than about 150 to 200, where
p is the gas pressure in millimetres of mercury and d the electrode separation in
centimetres. This means that important practical cases such as the electrical breakdown of a 1 cm gap in the atmosphere is not explained by currently accepted theory.
Electrical phenomena in gases at high values of pd are of considerable practical
interest owing to the increasing use of compressed gases for insulation in high-
voltage apparatus in nuclear physics research and in power engineering.

There are no reliable data concerning the ionization coefficients (representing
collision processes involving electrons, ions, photons, etc.) by which conflicting
theories can be tested when pd is large. For low values of pd, however, such that
the sparking potential in most gases is less than about 2000 V, a satisfactory and
generally accepted explanation of the static breakdown mechanism can be given
in terms of elementary processes such as ionization by collision with electrons,
photo-ionization, secondary emission from the cathode due to positive ions,
photons, excited atoms, or ionization by other more complicated processes. Under
these conditions both theory and experiment show that the growth of a small
initial photoelectric current I_0 from the cathode in a uniform electric field E may
be represented by the well-known Townsend relation

$$I = I_0 \exp(\alpha d)/[1 - (\omega/\alpha)\{\exp(\alpha d) - 1\}], \tag{1}$$

where α is the primary coefficient representing ionization by electrons, and (ω/α)
is a generalized secondary coefficient representing any or all of the secondary
processes. This representation is possible because the well-known secondary
processes, whether due to photons, ions or metastable atoms, all lead to an

*Reproduced from *Proceedings of the Royal Society,* **A213**, 185, (1952) by permission of The
Council of The Royal Society, London.

expression of the analytical form of equation (1). Any gaseous photo-process may be neglected at low pressures in comparison with the other secondary processes. On this theory the breakdown potential V_s of the gas is given by

$$0 = 1 - g(V_s/pd)[\exp\{f(V_s/pd)\,pd\} - 1], \tag{2}$$

and can therefore be calculated from data giving α as a function f of V_s/pd, and ω/α as a function g of V_s/pd and the cathode material. On the other hand, for sparking at high values of pd, while α has been determined in the usual way, previous work in this field has yielded no determinations of (ω/α) in spite of repeated attempts. The reason for this failure will be explained below. As a result, no satisfactory explanation of the mechanism of static breakdown at high values of pd has been formulated when that based on equations (1) and (2) has been rejected. Clearly, the α process alone in a uniform field is inadequate to account for breakdown. The conclusion that equations (1) and (2) are inapplicable at high values of pd in uniform fields has led to the introduction and wide acceptance of other theories known as streamer theories based on the assumption that the pre-breakdown ionization currents do not follow equation (1) (see Raether, Paper 3; Meek, Paper 5; Loeb and Meek, 1941). It is the purpose of this paper to show that this assumption is not in accordance with the facts, and that experiment shows that the static breakdown process in air at high values of pd, corresponding to a 1 cm spark in the atmosphere, is in fact in full agreement with the relations (1) and (2).

The paper describes investigations on the growth of small ionization currents involved in the pre-breakdown mechanism in air in uniform fields at high values of pd, in particular, at these values of E/p (39 to 45 V/cm/mm Hg) and of pd (≈ 800 mm Hg cm). There are many reasons, however, why air is not the most suitable gas to use when investigating the fundamental processes. Air is a mixture of gases with different ionization and excitation potentials, and the passage of high-energy electrons can cause the formation of oxides of nitrogen; but, above all, the main disadvantage is the oxidation of the cathode in the presence of air, and the consequent variation in the surface electrical properties (Llewellyn Jones, 1949). Nevertheless, air was the first gas to be investigated in these experiments, partly because previous experiments on measurements of the ionization coefficients at high values of pd have been carried out in air, but mainly because of the import-ance of that gas as a general insulator, whether in the open atmosphere or at high pressure in enclosed equipment. Further, the lightning flash is a natural example of the electrical breakdown of air at very high values of pd which still awaits full explanation.

Before describing the experiments, it is necessary to analyze the results of previous work on the growth of photoelectric currents at high pressures, as it was mainly on the results of such work that some present-day theories of breakdown have been based (Loeb, 1939).

Previous experimental data on ionization coefficients

Four experimental investigations (Paavola, 1929; Masch, 1932; Sanders, 1932; Hochberg and Sandberg, 1942) have been made of the breakdown of air at high values of pd. The methods employed were similar to that used originally by Townsend: the growth of an initial photoelectric current I_0 with electrode spacing d, was measured under uniform electric fields E at various values of E/p; log I was then plotted against d for constant values of E/p. The slope of the straight portion of the graph gave α; where the graph was not linear, (ω/α) could be determined by inserting known values of I, I_0, α and d in equation (1). It is necessary to consider this previous work in detail.

Paavola's results

   Paavola first attempted to measure the coefficients with an open-air spark-gap, but he found that consistent results could not be obtained, and this he attributed to the variable water content of the air. He later used an enclosed spark-gap with dry air at nearly atmospheric pressure, with electrode spacings up to 5 mm. He could not measure α when $E/p < 30$ V/cm/mm Hg. Further, he could not detect any departure from linearity of the graph of log I against d for the important region when $E/p = 40.4$ V/cm/mm Hg, which is the value corresponding to a spark in the atmosphere. At higher values of E/p he did, in fact, obtain values for (ω/α) which, however, he stated were not constant at a fixed E/p; i.e. the value of (ω/α) obtained by inserting the measured values of I, I_0 and d and the calculated value of α into equation (1) depended upon d. Consequently, it has been suggested (Loeb, 1939) that the non-linearity of the curves of log I against d for the larger values of d obtained by Paavola was, in fact, due to space-charge distortion because of the high current densities used in his investigations. The values of I_0 given in his graphs are $\approx 10^{-12}$ A, and the area of the cathode illuminated by ultra-violet light was 4 cm^2. Consequently, Paavola's data are unsatisfactory until further information is available.

Masch's results

   Masch used two ionization chambers; the first for $p < 760$ mm Hg was that used by Paavola; and the second was used for higher pressures. He measured the current I by determining the voltage drop across a resistor of known value (10^9 to 10^{11} Ω) in series with the spark gap. I_0 was of the order of 10^{-12} A, and the area of the cathode illuminated was 4 cm^2 in the Paavola spark gap, while the area was not given for the chamber used for high pressures. Masch measured α for air at pressures from 1 mm Hg to 2 atm for $500 > E/p > 31$, but no values of (ω/α) were recorded, and since no values of pressure are given for air, it is not possible to decide whether a secondary coefficient could have been detected from his measurements.

Sanders' results

   Sanders measured I in air contaminated with mercury vapour by measuring the potential to which a condenser C in series with the spark gap was charged in a

known time (10 s). The anode was disconnected from the voltage supply before
the condenser was connected to the Dolazalek electrometer used for measuring
the voltage across C in order to avoid induced potentials. The current I_0 was about
10^{-12} A, and the area illuminated was a rectangle 3 x 6 cm, and α was measured
in air at 380 mm Hg for $20 < E/p < 36.5$ V/cm/mm Hg. The most important
conclusion reached by Sanders was the fact that there was no discernible secondary
coefficient in his experiments, even when he was 'within 2% of sparking'. From
this result it has been widely concluded that no secondary ionization process occurs
in air when $E/p \approx 40$ (Loeb, 1939, pp. 371 and 407). If this were true, then it is
clear that high-pressure sparking requires some entirely new ionization mechanism,
which is not based on the relation (1). Consequently, it is necessary to examine
very closely the data upon which this assumption is based.

At $E/p = 36$ V/cm/mm Hg, Sanders found that α/p was 0.0082; and if this value
is inserted in the Townsend criterion for sparking, with $pd = 5000$ mm Hg cm
(Whitehead, 1926), it follows that $(\omega/\alpha) = \exp(-0.0082 \times 5000) = \exp(-41)$.
Now if
$$I_1 = I_0 \exp(\alpha d),$$
and
$$I_2 = I_0 \exp(\alpha d)/[1 - (\omega/\alpha)\{\exp(\alpha d) - 1\}],$$

then the fractional difference between these two values of gas-amplified current
$$(I_2 - I_1)/I_2 = (\omega/\alpha)\{\exp(\alpha d) - 1\}.$$

The possibility of measuring this fractional difference depends on the magnitude
of $(\omega/\alpha)\{\exp(\alpha d) - 1\}$, which must not then be negligible compared with unity.
In Sanders's experiments, when d was 2.5 cm, αd was 7.8, so that
$$(I_2 - I_1)/I_2 = \exp(-41) \exp(7.8)$$
$$= \exp(-33.2),$$

which was so small as to lie outside the possibility of measurement. Therefore, if a
secondary process were actually occurring in his experiments, Sanders would not
have been able to detect it under the conditions of his experiments because his
electrode separation was never close enough to the sparking distance. Consequently,
the fact that he was unable to record a value of (ω/α) is no evidence against its
existence.

Hochberg and Sandbergs' results
    The range $38 < E/p < 45$ V/cm/mm Hg was investigated at a pressure of
300 mm Hg, with a maximum electrode separation of 8 mm, i.e. pd was
240 mm Hg cm. The details of the methods employed to measure the current and
voltage are not given in their paper, but a circuit diagram indicates that the methods
of generating the voltage and its measurement are similar to those employed
previously by Paavola. The initial current I_0 was 10^{-12} A and the area illuminated
was 12 cm^2. Calculation shows that with $E/p = 45$ V/cm/mm Hg the sparking
distance at 300 mm Hg was 1.3 cm, and that I_1/I_2 for $d = 0.8$ cm was 99.5%, and

it is probable that this 0·5% change in current would be within the limits of their experimental error. Therefore it was not possible for Hochberg and Sandberg to detect a secondary coefficient in their experiments. The values of α/p given by Masch and those of Sanders agree while those of Hochberg and Sandberg and Paavola are also in fair agreement; there is no good agreement, however, between the two pairs of results. Therefore, no determinations of α/p exist confirmed by several independent workers. It is probable, also, that all results of the above experiments to date are for air contaminated with mercury, as the precautions taken to prevent contamination were not published.

Conclusion

In spite of the amount of work done previously on the high-pressure breakdown both theoretical and experimental, and in spite of widely accepted views to the contrary, no reliable data exist to show whether or not the growth of photoelectric currents at high values of pd follow the relation (1), i.e. whether the graphs of $\log I$ against d are linear or not right up to the sparking distance.

Consequently, experimental investigation of the growth of photoelectric current at high pd, and also the measurement of the ionization coefficients, must form an important first step in the investigation of the breakdown mechanism of gases at high pressures. The design, construction and operation of an ionization chamber with which such an investigation has been carried out will now be described.

Apparatus and procedure

Factors governing general procedure

Measurement of α and (ω/α) involves the determination of four parameters, viz, pressure, voltage, current and electrode spacing. The following description of the design of the apparatus shows how this is done; but the optimum experimental conditions of electric force, pressure, electrode separation and current density for measurement of α and (ω/α) will first be discussed. According to previous data (Whitehead, 1926) the value of V_s when $E/p = 40$ V/cm/mm Hg is 40 kV, and the corresponding value of pd is 1000 mm Hg cm; so that for an electrode separation of 5 cm the pressure must be 200 mm Hg. The effect of the secondary process (ω/α) can only be observed experimentally when $(I_2 - I_1)$ is greater than the experimental error in measuring I_2. To fix ideas, let this be 10% (previous published data indicate possible experimental error of this magnitude in the determinations of I); then the effect of a secondary process would not be observed unless

$$(I_2 - I_1)/I_2 = (\omega/\alpha)\{\exp(\alpha d) - 1\} = 0·1.$$

The values of the factor $(\omega/\alpha)\exp(\alpha d)$ for values of d up to 5 cm when $E = 8000$ V/cm and $p = 200$ mm Hg, are given in *Table 1*.

It is clear that, under these conditions, the influence of the factor (ω/α) can only reach a measurable value over the comparatively short range of electrode separation from 4·3 to 5·0 cm. Thus, any secondary effect does not become significant until

Table 1 – Values of (ω/α) exp (αd) for Various Values of d, at (E/p) = 40 V/cm/mm Hg and p = 200 mm Hg. Sanders' Values of (α/p) are used

electrode separation d (cm)	2	3	4	4·5	4·6	4·7	4·8	4·9	5
(ω/α) exp (αd)	–	$1·23^3 \times 10^{-3}$	$3·4 \times 10^{-2}$	0·18	0·25	0·35	0·49	0·68	1·0 (spark)

gas-amplified currents are measured at electrode separations which are within 20% of the sparking distance.

The ionization chamber

This should, if possible, satisfy the following conditions:

(i) A uniform field must exist between the electrodes throughout the region where the electron stream moves; thus the electrode diameter D must be greater than three times the maximum electrode separation.

(ii) It must be possible to determine the separation d to the same high accuracy as the pressure measurement.

(iii) The electrode which received the measured ionization current must be at low (earth) potential; also, it must be highly insulated.

(iv) It is an advantage to have the low potential electrode the movable one, and to have the electrode which receives the measured current as small as possible, in order to reduce the chance of picking up stray charges.

(v) Pressures up to 1 atm are to be investigated initially, but in later work now being planned, pressures up to 3 atm will be investigated; hence the ionization chamber was designed to withstand these high pressures.

(vi) It should be possible to view the interior of the chamber in order to observe sparks.

(vii) The chamber should if necessary be able to withstand temperatures of more than 100° C, in order to assist the driving off of absorbed gas, particularly water vapour.

(viii) Screw gauges for measuring the electrode separation should be external to the envelope and easily detachable, in order to be unaffected when the apparatus is heated.

(ix) The source of ultra-violet light should preferably be at the low potential end to avoid the necessity of electrical insulation.

All the above desirable conditions cannot be satisfied together; and in practice condition (ix) was relaxed in order to permit the cathode to be at earth potential.

The design of the ionization chamber, which was constructed in the laboratory, is shown diagrammatically in *Figure 1*. A large boro-silicate glass flange F with ends ground flat formed a basis for the envelope. This was mounted on a mild steel base-plate P, and the seal made vacuum-tight by a rubber gasket G, coated on each side with a thin layer of silicone grease. The cathode C was insulated by means of three quartz rods and mounted on a cylinder S which screwed into the base-plate. The electrode separation was varied by turning S externally, and electrical contact

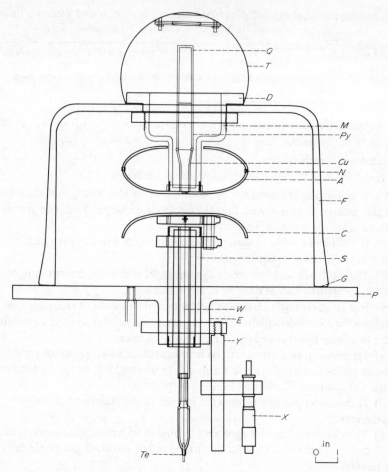

Figure 1. Ionization chamber.

was maintained with *C* by a wire *W* through the seal *Te* in a glass tube which was itself sealed to a copper tube. The anode *A* was screwed on a mild steel cylinder *M*. The upper flange *D* of the anode support *M* rested on the top of *F*, and the seal was made vacuum-tight by a silicone-grease-coated rubber gasket. Ultra-violet light was admitted into the chamber through a quartz window inside the spherical terminal *T*, and the window *Q* was joined by means of a graded seal to a Pyrex tube *Py* which was joined to a copper tube *Cu*, silver-soldered into the base of cylinder *M*.

The maximum diameter of electrodes which could be constructed from available pure nickel sheet was 15 cm, so the maximum electrode separation which could be employed was 5 cm. Electrodes with a Rogowski profile were used; and an investigation of a full-scale cross-section of the electrode system in an electrolytic bath showed that the field was uniform between the important central plane portions at all electrode separations. As the ultra-violet light fell on an area of the cathode of only 0·4 cm radius, the motion of photoelectrons took place in a uniform field. The electrodes were hollow so that their mass would be as low as possible to facilitate out-gassing, and were constructed from 3/16 in. sheet nickel. Nineteen holes, each 1 mm diameter, were drilled in the centre of the anode, inside a circle of 8 mm diameter, to admit the ultra-violet light to the centre of the cathode. The plane portions of the electrodes could be made parallel by adjusting the quartz supports. The electrodes were buffed, and the plane portions were rubbed with emery paper, finishing with the 0000 grade and finally polishing with alumina.

The electrode separation was measured by means of a micrometer X supported by a mild steel framework Y fastened to the portion E of the base-plate. The cathode was raised until it made electrical contact with the anode, and the micrometer was then screwed up to touch the base of the cylinder S; this was taken as zero separation. The micrometer scale was then set to the required electrode separation, and the cylinder S unscrewed until its base made contact with the micrometer shaft. The micrometer could be read to 1/200 mm so that changes of 0·5 mm in the electrode separation were the smallest which could be made with sufficient accuracy. The framework E was easily removable, so that no damage to the micrometer would occur in any outgassing processes.

Gas system

Great care was taken to exclude mercury vapour from the ionization chamber and the gas system. An oil-diffusion pump was used to evacuate the chamber, and low pressures were measured with a Pirani gauge. High pressures were measured by a specially designed mercury manometer from which mercury could not diffuse. The column bore was 3 mm while that of the reservoir was 100 mm, so that the movement of the mercury surface there was less than 0·5 mm. This mercury surface was covered with a layer of silicone vacuum oil 15 mm thick, and as a further precaution against the diffusion of mercury vapour the manometer was separated from the gas system by a trap containing tin foil. No movement of gas from the manometer to the system was allowed; flow was always from chamber to manometer from which it could be evacuated by a subsidiary pump. The air used in the experiments was carefully filtered and dried over phosphorus pentoxide.

Voltage supply and ionization current measurement

Since α increases very rapidly with the field, it was most important that the voltage supply should be extremely steady. The mains were unreliable because of switching surges, so the power supply was taken from an ex-Admiralty 500 Hz motor-alternator fitted with a carbon-pile stabilizer and driven by large accumulators. The circuit used was a voltage doubler employing two rectifiers rated at 125 kV peak inverse and fed from a high-voltage transformer through a Variac. The smoothing circuit consisted of a bank of ten 0·25 μF condensers in series, centre-tapped for voltage doubling, and followed by a filter circuit consisting of a 1 MΩ resistor and a bank of four 0·5 μF condensers in series. Calculation showed that any ripple was negligible ($< 10^{-4}\%$), and, in fact, no ripple could be detected.

On the other hand, slow fluctuations of voltages of approximately $4 \times 10^{-2}\%$ still remained, being the limit of stability of the motor-generator and battery. The power supply was connected to the anode through a 5 MΩ resistor to limit the current in the event of sparks passing.

Corona losses, which caused variations in the anode potential, were eliminated by careful distribution of electric stresses over the glass envelope F with colloidal graphite, covering sharp edges with plasticene, and also by using large spherical terminals.

The most direct method of measuring the anode potential was to measure the current through a high resistance in parallel with the source, but to avoid excessive potential drop across the smoothing resistor the current drain had to be small, say 1 mA, when the anode potential was 100 kV. A study was made of the small variation in the resistance of the cracked carbon-type resistor under load, and the cause was found to be the rise of temperature due to Joule heating; variation of resistance was eliminated by immersing the resistor in circulating oil. The completed high resistor consisted of a chain of a hundred 1 MΩ cracked carbon resistors mounted in a glass tube in the form of a large W through which oil was pumped. During use the change of resistance was approximately $10^{-3}\%$. The voltage (≈ 300 V) developed across the last resistor in the chain at the earth end was balanced against an equal voltage produced by a subsidiary stabilized source connected across a standard resistor acting as a potentiometer; a known small fraction of the voltage was balanced against a standard cell. Thus the total high potential of the anode was given directly in terms of the voltage of the standard cell.

The ionization currents were measured by a screened electrostatic balance and a Dolezalek electrometer, by which the cathode was maintained at earth potential while the currents passed. Unavoidable electrometer drift set a lower limit ($\simeq 5 \times 10^{-14}$ A) to the current which could be measured to the required accuracy by this electrometer system, but the practical lower limit to the ionization current which could be measured was set, not by this, but by the degree of stability of the power supply. Any variation of anode potential V was distributed between the capacity of the electrodes (10 cm) and that of the electrostatic balance (200 cm). A variation of 0·04% when V was 30 kV meant that the potential of the anode in the chamber changed by 12 V, which consequently produced a change of 0·6 V on the electrometer. This change, occurring over the maximum time (120 s) taken to make a current measurement, was equivalent to a current of 10^{-12} A, and this represented the limit to the current which could be measured in this way.

This limitation had an important bearing on the range of electrode spacings at which the ionization currents could be measured. It is usually considered that space-charge distortion is likely with currents $> 10^{-8}$ A. Hence, in order that the gas-amplified currents at the large electrode spacings near the sparking distance should lie within the measurable range 10^{-12}-10^{-9} A, it was necessary to employ very low initial photoelectric currents of 10^{-15} A. Currents in the range 10^{-15}-10^{-12} A, obtained with the lower range of electrode spacings, could not then be measured with the same apparatus as that used for the higher amplified currents in the vicinity of the sparking distance. This, however, was no great disadvantage, since I_0 (= I exp $(-\alpha d)$) could be calculated from the value of the currents I obtained at the larger spacings in the range when the graph of log I against d was still linear. Only a rough check on the value of I_0 was possible by using anode potentials of about 100 V obtained from accumulators and measuring I_0 when the chamber was evacuated.

The initial photo-current

Tests with a photo-cell showed that the intensity of the ultra-violet light from the mercury arc was constant within the required limits, but measurements of the initial photo-current I_0 from the cathode in the ionization chamber was steady to within about 3% only over a period of some hours. Clearly this was insufficient.

Since it was of the greatest importance that I_0 should be constant throughout any given set of measurements of I at various electrode separations d, considerable attention was paid to the state of the cathode surface and the constancy of I_0 at all stages of the measurements. Although over a period of many weeks I_0 gradually changed, it was found that when the cathode had been cleaned and polished and only small currents passed, still less any sparking permitted, I_0 remained constant within the required limits during experiments lasting several days. *In vacuo*, I_0 was independent of the anode potential, and with gas pressures up to 200 mm Hg, I_0 was independent of the electrode separation indicating that absorption of the ultra-violet light by the air in the chamber was negligible.

When d exceeded 3·0 cm, the values of I during any series of measurements increased progressively. In a typical set when d was 3·0 cm, I increased from $1·14 \times 10^{-10}$ to $1·82 \times 10^{-10}$ A in a set of seven determinations at 30 s intervals. Experiment showed that such increases depended only on the previous passage of large ionization currents; but no current from the cathode could be detected unless the cathode was being irradiated, showing that no effects due to a delayed cathode emission were taking place (Paetov, 1939; Llewellyn Jones, 1949). When d was 3·5 cm, so that higher gas-amplified currents were passing, I increased from $1·5 \times 10^{-10}$ to $4·0 \times 10^{-10}$ A after the maintenance of the current for 5 min. Clearly this effect, which will be discussed further below, was due to alteration of the cathode surface produced by the high gas-amplified ionization currents produced when d was near the sparking distance, and this effect was one of the difficulties inherent in the use of air which caused progressive change by the oxidation of the cathode under the action of positive ion bombardment. On the other hand, it was found that if only three successive determinations of the current were made at any given value of d in a time of about 1·5 min, and if an interval of about 10 min elapsed between measurements at different electrode spacings, then the general increase of I with successive determinations did not occur, and repeatable determinations of the currents could be made at all distances up to the sparking distance.

Currents were measured at 200 mm Hg air pressure for electrode spacings from 1 to 3·8 cm in steps of 1 mm, using values of anode potential up to 30 kV, of pd up to 760 cm mm Hg, and of E/p from 39–45 V/cm/mm Hg.

Results

Typical results are given in *Figure 2*, in which $\log I$ is plotted against d. Two conclusions clearly follow:

(i) when d has low values, the $\log I - d$ relationships are linear, in agreement with all previous investigations in this field; and

(ii) when d approaches the sparking distance d_s distinct departures from linearity occur, the current increasing faster than exponentially. These curved sections near the sparking distances lie well outside the experimental spread in the determinations of the current at any given electrode spacing.*

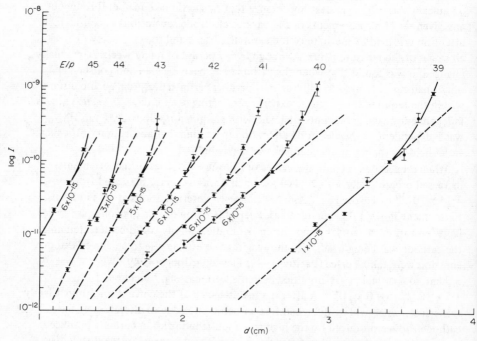

Figure 2. Log $I - d$ relations for air for values of E/p from 39 to 45 V/cm/mm Hg. Values of I_0 are indicated on the curves.

Analysis of these graphs consists of finding whether they can be represented by equation (1) for certain values of I_0, α and (ω/α) at any given value of E/p. When the secondary ionization coefficient (ω/α) is small (say $< 10^{-4}$), the term $(\omega/\alpha) \exp (\alpha d)$ is negligible compared with unity at the smaller values of d, and

* An advance notice of this result was published in *Nature* (Llewellyn Jones and Parker, 1950).

then the $\log I - d$ graph is a straight line of slope α and the intercept on the axis is $\log I_0$. The graphs of *Figure 2* can be represented very well by equation (1) for the values of the constants and parameter E/p given in *Table 2*. The curve for E/p = 45 was too short for accurate analysis; in this case therefore Masch's value for α was assumed and I_0 was taken to be the same as that found in measurements with E/p = 40 made just before those for E/p = 45.

Table 2– Experimental Values of I_0, α/p, (ω/α) and Calculated Values of d_s for Various Values of E/p when p = 200 mm Hg

E/p (V/cm/mm Hg)	$10^{15} I_0$ (A)	α/p (cm mm Hg)$^{-1}$	$10^6 (\omega/\alpha)$	d_s (cm)
39	1	0·0161	8	3·73
40	6	0·0181	23	2·94
40	20	0·0181	$\geqslant$ 150	–
41	6	0·0196	40	2·53
42	6	0·0224	46	2·22
43	5	0·0252	113	1·83
44	3	0·0295	105	1·57
45	6	0·0345	84	1·37

The sparking criterion

When the growth of the initial current I_0 follows equation (1) the electrode spacing d_s at which the current becomes self-maintained and independent of I_0 is given by

$$d_s = \alpha^{-1} \log (1/(\omega/\alpha)),$$

and from this the sparking potential $V_s(= Ed_s)$ can be obtained. For reasons given above, sparking was permitted only very infrequently and only when values of V_s were required; this was to avoid unduly affecting the cathode surface by the spark current. When the cathode surface was stable, and at the termination of a set of current measurements, the electrode spacing was gradually increased until a spark occurred; the potential was then noted. The theoretical and observed values of the sparking potential thus obtained are plotted against pd in *Figure 3*; it can be seen that within the experimental error they are exactly the same. In the measurements at E/p = 40 with $I = 20 \times 10^{-15}$ A the cathode was becoming unsteady as d increased, so V_s was not measured.

Significance of the results

The most important result of these experiments is the support they furnish to the view that the original Townsend-type equation (1) for the growth of the photo-electric currents holds for values of E/p in air in the range 39 to 45 V/cm/mm Hg; this is a range which corresponds to sparking potentials from 12 to 30 kV and values of pd up to 760 mm Hg cm. This fact has never previously been established.

This view (that there is no secondary ionization coefficient (Loeb and Meek, 1941)) is not confirmed by the present experiments; on the contrary, a clearly defined departure from linearity of the $\log_e I - d$ curves occurred when the voltage

Spark Mechanism in Air

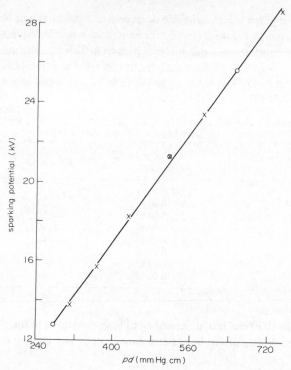

Figure 3. Calculated (×) and observed (○) sparking potentials
in air.

across the gap was within 2% of the sparking potential. The failure by Sanders and others to observe a secondary effect at these values of E/p was due to the fact that no measurements of the currents I were made at voltages which were less than 2% below the sparking potential, as the analysis, given above, has indicated. With the apparatus used in the present investigation, currents could easily be measured at voltages which were only 0·5% below the sparking potential. From present measurements given in *Figure 2* it is seen that no departure from linearity occurred in the $\log I - d$ curve until the electrode separation reached at least 80% of the sparking distance. *Figure 3* shows that the observed values of sparking potentials agree with those calculated from equation (2) well within the limits of experimental error.

The significance of these conclusions may be seen by referring to the discussion of current theories of spark breakdown. The two theories of breakdown (viz. the theory based on the rapid development of currents by primary and secondary ionization processes, which is generally accepted for low values of the parameter pd less than about 200 mm Hg cm, and 'streamer' theories of Raether and Loeb and Meek, based on the sudden introduction of space-charge phenomena and photo-ionization of the gas, and proposed for the higher values of pd greater than about 200 mm Hg cm) must now be re-examined in the light of the present results.

The absence of any experimental values for a secondary coefficient (ω/α) was one of the main reasons which led to the formulation of the streamer theory of static breakdown, but the present experimental results show that Townsend's equation does, in fact, hold for electrode separations, right up to sparking under static electric fields at comparatively high values of the product pd (≈ 760 mm Hg cm), i.e. well within the range for which it has been assumed that the Townsend theory fails and the 'streamer' theory is necessary. Further, in this range of pd the Townsend theory can predict values of sparking potential from the values of the ionization coefficients obtained from measurements on pre-breakdown currents, a procedure which is not possible on the present streamer theories. This fact supports the view that the actual electrical breakdown of the gas is a continuation of the process of the growth of the pre-breakdown ionization currents according to the relation (1).

Nature of the secondary ionization in air at high pressures

It is of interest to consider whether any conclusions can be drawn from the present work about the nature of the secondary ionization processes (Thomson and Thomson, 1933). Evidence that the secondary ionization observed in the present work must in part be due to secondary emission from the cathode is furnished by the following considerations. The constancy of the light emitted by the mercury-arc source was established by regular measurements with a photo-electric cell during the experiments; but *Table 2* shows that, in these experiments, two values of I_0, viz. 6×10^{-15} and $2 \cdot 0 \times 10^{-14}$ A, were obtained by photo-emission at the cathode when E/p was 40 V/cm/mm Hg, although the incident light intensity remained constant; this change in I_0 was doubtless due to a change in the nature of the cathode surface which occurred during the course of the experiments. The fact that the photoelectric properties of the cathode surface could be changed by the passage of a small current between the electrodes was demonstrated during the experiments on the photography of the spark discharge; after the passage of a low-current ($\approx 20 \mu$A) discharge for some minutes during the course of those experiments, it was noticed that the photoelectric sensitivity vanished, and examination showed that the cathode was covered with a very thin oxide film (interference fringes were clearly visible). Removal of the film from the cathode surface restored its photoelectron emissive properties, so that it is reasonable to suppose that the presence of this oxide film was the cause of the photoelectric insensitivity. The continual formation of oxide layers on the nickel cathode is an unavoidable consequence of the use of air in these experiments; positive ions of oxygen react with the cathode to form such oxide layers, and this is a great disadvantage in any study of cathode ionization processes in the spark discharge. For any such study a more inactive gas, such as nitrogen, or a monatomic gas, is desirable. Nevertheless, the fullest possible information on the spark mechanism in air is necessary, as pointed out above, not only because most published data on spark breakdown at high pressures have been obtained with air, but also because of the technological applications of that gas as an insulator.

Consider now the values of the secondary coefficient (ω/α) obtained from the growth of a photoelectric current in two cases when there was reason to believe that the nature of the cathode surface had changed. *Table 2* shows that when E/p was 40 V/cm/mm Hg, I_0 was 6×10^{-15} A, (ω/α) was $2\cdot3 \times 10^{-5}$; and it can also be seen that when I_0 had increased to $2\cdot0 \times 10^{-14}$ A (due to a change of the photo-emissive properties of the cathode surface), (ω/α) had also increased to more than $1\cdot5 \times 10^{-4}$. Thus, when the number of photoelectrons produced per second at the cathode surface by a source of constant intensity was increased (because of a change in the nature of the surface), the value of the secondary coefficient also increased. This result is evidence for the view that the secondary effect (ω/α) is in part due to secondary emission from the cathode (photoelectric effect and/or positive-ion bombardment).

It is now necessary to discuss the important result of the preliminary experiments described above in the section on apparatus and procedure, because its interpretation confirms the view that the secondary effect is in part due to secondary emission from the cathode. Those preliminary results showed that for values of electrode separation near the sparking distance, the continual passage of a gas-amplified current resulted in a continual increase in I. It is important to note that such a gradual increase of I with passage of current (the other factors E/p and d remaining constant) was not observed for electrode separations less than about 70% of the sparking distance when the total current I was low ($\approx 10^{-12}$ A), the gradual increase only occurring when the total current I was large ($\approx 10^{-9}$ A) and a pronounced secondary ionization could be detected by the upcurving of the log I graph. It is unlikely that the incidence of radiation would greatly affect the cathode surface, but, on the contrary, it is easy to see that positive ions, especially of oxygen, can produce profound changes in the cathode surface owing to the formation of thin oxide films. It is significant, therefore, that the increase observed in I took place only when a pronounced cathode emission was also taking place. If, as seems reasonable, the change in the cathode surface was produced by the action of positive ions, it is now necessary to consider how such a change could increase the gas-amplified current I.

It has been shown that the above passage of even small pre-breakdown currents in air leads to the formation of thin oxide layers on the cathode, and the electron emissive properties of such oxide films in discharges have been studied in this laboratory (Llewellyn Jones, 1949; Llewellyn Jones and Morgan, 1951). It has been established that greatly enhanced electron emission from the cathode under an electric field can be produced in the presence of thin films. The mechanism of the process, briefly, is that positive ions are swept to the cathode, fall on the outer surface of the thin partly insulating oxide film, and thus set up an intense local electric field at the cathode surface. This field may extract electrons from the metal, or provide electrons at the oxide surface from which they would easily be removed by bombardment by other positive ions. Thus, enhanced electron emission takes place as a result of positive-ion bombardment. The net result is, therefore, an increase in the gas-amplified current I owing to an effective increase in (ω/α). Such an increase in current could only be observed in that section of the current-distance curve where the secondary effect contributes significantly to the total current, i.e. near the sparking distance. This is just what was observed. Thus the formation of a thin oxide layer on the cathode can affect the measured current in two ways. First, it can produce a *reduction* in I_0 by reducing the photoelectric efficiency of the cathode surface; and, secondly, it can produce an *increase* in the gas-amplified current at electrode separations near to the sparking distance owing to enhanced secondary electron emission from the cathode, this enhanced emission being due to the positive charge acquired by the oxide surface. There is this difference, however; the change in I_0 (due to a change in cathode surface) is gradual and semi-permanent, while the effective increase in (ω/α) is instantaneous and only occurs when the positive-ion current flowing into the cathode is large, just as was observed.

The present experiments in air give no direct information about the relative magnitude of the two cathode secondary processes; but a rapid variation of (ω/α) with E/p is usually taken to indicate the presence of photo-electric emission due to the incidence of photons from the electron avalanche (Llewellyn Jones and Davies, 1951). The values of (ω/α) given in *Table 2* are in accordance with the view that some photoelectric emission, as well as emission due to the bombardment by positive ions, played some part in the secondary ionization processes represented by the coefficient (ω/α).

Conclusions

The most important conclusion derived from the present results is that the pre-breakdown growth of ionization currents in air in uniform electric fields can be represented by the relation (1) in the range of spark parameters which corresponds to sparking potentials between 12 and 30 kV and values of pd from 300 to 760 mm Hg cm.

The results of previous measurements of gas-amplified currents at high pressures have been misinterpreted; it has been widely assumed that because no departure from linearity of the $\log I - d$ curves had been obtained in previous measurements no secondary ionization process existed to amplify the pre-breakdown current and lead to the spark.

The agreement in the present experiments between calculated and observed sparking potentials is consistent with the view that static spark breakdown at high values of the parameter pd is brought about by the same mechanisms as amplify the pre-breakdown current, i.e. by primary and secondary ionization processes in uniform fields.

The experiments support the view that in uniform fields cathode emission plays an important part in the secondary ionization process, and that in air it is probable that emission is produced both by the incidence of photons and of positive ions. Further elucidation of this question requires the investigation of the growth of currents in pure non-active gases.

References

Hochberg, B. and Sandberg, E., *J. Tech. Phys. U.S.S.R.* **12**, 65, (1942).

Llewellyn Jones, F., *Proc. Phys. Soc.* B, **62**, 366, (1949).

Llewellyn Jones, F. and Davies, D. E., *Proc. Phys. Soc.* B. **64**, 519, (1951).

Llewellyn Jones, F. and Morgan, C. G., *Phys. Rev.*, **82**, 970, (1951).

Llewellyn Jones, F. and Parker, A. B., *Nature, Lond.*, **165**, 960, (1950).

Loeb, L. B., *Fundamental Processes of Electrical Discharges in Gases*, pp. 371, 386, 407, New York: John Wiley, (1939).

Loeb, L. B. and Meek, J. M., *The Mechanism of the Electric Spark.* Stanford: The University Press, (1941).

Masch, K., *Arch. Elektrotech.* **26**, 589, (1932).

Meek, J. M., *Phys. Rev.* **57**, 722, (1940).

Paavola, M., *Arch. Elecktrotech.* **22**, 443, (1929).

Paetov, H., *Z. Phys.* **110**, 770, (1939).

Raether, H., *Z. Phys.* **112**, 464, (1939).

Sanders, F. H., *Phys. Rev.* **41**, 667, (1932).

Whitehead, S., *Dielectric Phenomena in Gases.* London: Benn, (1926).

Introduction to Paper 7 (Fisher and Bederson, 'Formative time lags of spark breakdown in air in uniform fields at low overvoltages')

After the pioneering work of Rogowski (paper 2) and others, a number of authors studied the so-called 'temporal growth' of ionisation. In particular, the time lag which occurs between the application of a voltage greater than the breakdown voltage and the occurrence of a spark was examined.

It is necessary to distinguish between two classes of time lags, (a) statistical time lags and (b) formative time lags. The former are caused by the delay which may occur before an initiatory electron appears in the discharge gap while the latter correspond to the time taken by the discharge to develop after initiation. While statistical time lags are obviously variable, formative time lags are calculable for given discharge conditions (see papers 8, 9 and 10).

At the time of the paper by Fisher and Bederson it was widely believed that the Townsend mechanism of breakdown held for low values of the parameter pd but that for high values of pd this was replaced by the streamer mechanism. Fisher (*Phys. Rev.* **69**, 530, (1946)) and Loeb (*Proc. Phys. Soc.* **60**, 561, (1948)) had, in fact, suggested that the transition depended not on pd but on p and d separately. Fisher and his colleagues attempted to examine the transition region and chose to do so by studying formative time lags for low overvoltages, i.e. by applying impulse voltages which were fractions of a percent greater than the breakdown voltage. Although Fisher and Bederson state in the paper given here that their data were inadequately explained by the Townsend theory, their remarks apply when the secondary Townsend mechanism is taken as being the release of electrons from the cathode by bombardment of positive ions created in the primary avalanche. They suggest in the last section of the paper that the observed time lags were more accurately predicted by assuming that the secondary electrons were produced from the cathode by photons released in the primary avalanche and in fact the analysis of Dutton, Haydon, and Llewellyn Jones (*Brit. J. Appl. Phys.* **4**, 170, (1953)) confirmed this view. Later measurements by Kachickas and Fisher in other gases confirmed that by suitable choice of the nature and magnitudes of the secondary mechanisms the Townsend theory could be used to predict time lags in agreement with experiment.

Further Reading

Kachickas, G. A. and Fisher, L. H., *Phys. Rev.* **88**, 878, (1952).
Köhrmann, W., *Z. Angew. Phys.* **7**, 183, (1955) and *Ann. Phys.* **18**, 379, (1956).
Aked, A., Bruce, F. M., and Tedford, D. J., *Brit. J. Appl. Phys.* **6**, 233, (1955).

7

Formative time lags of spark breakdown in air in uniform fields at low overvoltages*

L. H. FISHER AND B. BEDERSON†

1. Introduction

The theory of spark breakdown has undergone radical changes in the last ten years.[1-4] Until 1946 it was commonly believed that in air,[5] the Townsend mechanism of breakdown is valid below values of the product of pressure[6] p and plate separation d of 200 mm x cm, and that above this value of pd, the breakdown proceeds by the streamer mechanism. However,[7] on the basis of an analysis of Meek's equation for calculating sparking potentials, it was pointed out that the transition of Townsend to streamer mechanism may depend not on the product pd but rather on the values of p and d separately. It appeared impossible to study the transition by means of sparking potentials. A more hopeful approach seemed to lie in the study of formative time lags of spark breakdown.[8, 9]

The formative time lag is defined as the time necessary for a potential difference to be maintained across a gap before it breaks down, provided a primary source of ionization is present. From mobility measurements it is known that at the sparking potential the positive ion requires about $18\ \mu s$ to cross a one-centimetre gap at atmospheric pressure. In 1936, Schade[10] measured formative time lags for the glow discharge in neon at very low pressures. He found time lags between $10\ \mu s$ and $0{\cdot}1\ s$. Schade could not measure times shorter than $10\ \mu s$ with his equipment, a circumstance of great importance in the subsequent theoretical development of the field. But even earlier, and continuing up to the present[11] there had been observed a formative time lag of spark breakdown in air near atmospheric pressure so short that the positive ions formed in the gap could not possibly have had time to cross the gap. These times have been found to be of the order of $0{\cdot}1\ \mu s$, some observers reporting formative times as short as $10^{-3}\ \mu s$. Thus, the Townsend theory was shown to be inadequate at pressures near atmospheric and the streamer theory of

* For preliminary reports of this work, see L. H. Fisher and B. Bederson, Brookhaven Gas Discharge Conference, October, 1948, *Phys. Rev.* **75**, 1324, 1615, (1949), Pittsburgh Gaseous Electronics Conference, November, 1949, *Phys. Rev.* **78**, 331, (1950).

† Reproduced from *The Physical Review*, **81**, 109 (1951) by permission of The American Physical Society, New York.

Loeb and Meek,[2,3] and Raether[4] was developed for this pressure region. The general impression, with the enormous prestige of the Townsend theory and the confirming work of Schade, was that the Townsend theory applies at small values of pd (where the time lags were found to be long), and that the streamer theory applies at large pd (where the time lags were found to be short).

The work reported here was undertaken to determine the transition region.

2. Apparatus and experimental procedure

The ionization chamber used in this study was previously employed by Sanders[12] and Posin[13] for measurements of the first Townsend coefficient in air and nitrogen respectively. The chamber was later modified to its present form[14] when it was used to measure sparking potentials in air. The chamber was originally constructed with the aid of a grant in support of Sanders's research by the National Research Council. A description of the chamber may be found in the original papers.[12,14] The levelling and polishing of the electrodes, the measurement of electrode separation, the admitting and drying of the air, and the measurement of pressure were carried out as described previously.[14] All pressures have been corrected to 22°C.

In order to avoid the measurement of statistical time lags, the cathode was illuminated with ultraviolet light from a d.c. quartz mercury arc. The light was focused by a quartz lens and passed through a quartz window in the side of the chamber. To determine the photoelectric current i_0 emitted at the cathode, the gas amplified current i was measured with a voltage across the gap of about 10 percent below the sparking potential V_s. An electrometer was used to measure the resulting potential drop V across a resistor R inserted in series with the gap. Thus $V = i_0 \exp(\alpha d) R$, where α is the first Townsend coefficient as given by Sanders.[12] Hence the number of electrons n_0 leaving the cathode per second is given by

$$n_0 = V/\{1 \cdot 6 \times 10^{-19} R \exp(\alpha d)\}, \tag{1}$$

where all electrical quantities are in practical units. The irradiation was adjusted so that at least several electrons were emitted each microsecond, and was large enough in nearly all cases to prevent statistical variations in the time lags yet small enough to prevent distortion of the electric field. The ultraviolet lamp was on continuously during a set of measurements. No elaborate provisions were made for stabilizing the d.c. voltage for the mercury arc, since the experimental data were found to be very insensitive to small changes in n_0.

In order to determine the formative time lags, an approach voltage $V_0 (V_0 < V_s)$ is applied across the gap. Then, at some time, an additional voltage V_1 (the pulse voltage) is applied, where $V_0 + V_1 > V_s$. The time that $V_0 + V_1$ must be maintained before the gap breaks down is the formative time lag.

The circuit diagram of the electronic apparatus is shown in *Figure 1*. The values of the circuit elements are given in *Table 1*.

Table 1–Resistances in Megohms and Capacitances
in Microfarads (unless otherwise noted)

R_1	$0.7\ \Omega$	R_{10}	20
R_2	0.1	R_{11}	$2000\ \Omega$
R_3	1	R_{12}	$25\ \Omega$
R_4	20	R_{13}	1
R_5	1	R_{14}	1
R_6	100	R_{15}	1
R_7	$1000\ \Omega$	R_{16}	1
R_8	$1000\ \Omega$	R_{17}	1
R_9	$3300\ \Omega$		
C_1	0.25	C_7	0.01
C_2	0.25	C_8	0.01
C_3	0.25	C_9	$50\ pF$
C_4	0.25	C_{10}	16
C_5	0.25	C_{11}	$1000\ pF$
C_6	0.25	C_{12}	$1000\ pF$

The V_0 power supply is shown in the upper section of the diagram. V_0 is measured by R_6, a bank of five 20-MΩ high accuracy wire wound resistors. The absolute value of R_6 is not known, but comparison of sparking potential measurements with those made at Berkeley[14] in the same chamber with precision Taylor resistors indicates that the value of R_6 is known to within one percent. (The

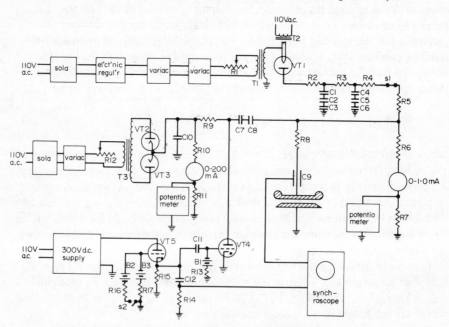

Figure 1. Circuit diagram. Values of circuit elements are given in
Table 1.

present experiment requires relative rather than absolute accuracy.) The ripple in V_0 was measured as 0·1 V at 20 kV. The voltage regulation is satisfactory, the voltage remaining fixed to within several volts over many minutes.

The V_1 power supply (shown in the centre section of *Figure 1*) is capable of delivering from 0 to 5 kV, with a maximum ripple of about 0·2 V. The trigger circuit is shown in the bottom section of the diagram.

The delay between the application of the sweep trigger pulse and the start of the sweep is about 0·1 μs. When VT_5 fires, VT_4 also fires forcing the potential of the high voltage electrode of the spark gap to fall by an amount equal to the value of V_1 minus the voltage drop in VT_4 while firing. The voltage across the gap becomes essentially equal to $V_0 + V_1$ in a time determined by the time constant of R_8 and the gap capacity. The capacity of the gap (with the low voltage electrode connected to the chamber) has been measured as 190 pF at $d = 1$ cm. Thus, the time constant for the rise of the pulse (assuming VT_4 breaks down instantaneously) is 0·19 μs; with $d = 0·3$ cm (the smallest electrode separation studied) the time constant is 0·4 μs. Actually, the time of breakdown of VT_4 is of the order of 0·5 μs, so that the time for the pulse to reach its full value across the electrodes ranges from about 0·6 to 1·0 μs, depending on electrode separation. Because of the long times observed in the experiment this rise time can be neglected. The time constant of the pulse decay circuit is 0·2 s. Since the present equipment was not used to measure times longer than 100 μsec, the pulse decay time is unimportant.

A signal from the pulse across the chamber is picked up on the vertical deflection plates of the synchroscope. The delay of several tenths of a microsecond in the breakdown of VT_4 is always sufficient to make the start of the pulse visible. When the spark gap breaks down, a sharp break is seen on the synchroscope trace. Normally, the spark signal has too high an amplitude to be seen. A typical trace is shown in *Figure 2*. The dotted part of the signal is due to the spark current and can be seen only when V_s is small.

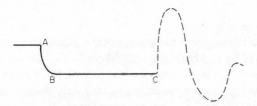

Figure 2. Typical synchroscope trace. A–Start of pulse; AB–pulse rise; C–beginning of spark; BC–time lag in inches.

The synchroscope has four sweeps; by using the various sweeps, times from about 0·5 μs to 100 μs can be measured. The sweeps were calibrated against a crystal oscillator.

After the chamber was filled with air, approximately fifty sparks were passed with the V_0 supply in order to season the plates.[8,14] After these sparks passed, V_s

assumed a value which at any given time was very definite, but which increased slightly, perhaps by a few tenths of one percent, as the measurements were carried out. This small gradual rise is probably due to the warming up of R_6. At any given time, V_s was reproducible to within 2 or 3 V.

After determination of V_s, a value of V_0 exactly 2 kV below V_s was applied. As V_s changed, V_0 was adjusted so as to maintain $V_s - V_0$ at 2 kV. (This procedure was justified by the consistent results obtained.) V_1 was then applied, and the time lag was read by visual observation.

The overvoltage ΔV is defined by $\Delta V = V_0 + V_1 - V_s - V'$, where V' is a correction due to the loss of part of V_1 in the circuit. The percent o.v. is defined as $100 \, \Delta V/V_s$.

At a selected value of ΔV and p, about 10 measurements of the formative time lag were made. These measurements constitute a run. Runs were made for various percent o.v. from about two to as close to zero as possible. The entire procedure thus far described constitutes a series, and such series were carried out at various pressures and plate distances. Measurements were then carried out to study the effect of increasing the ultraviolet illumination; the effect of changing V_0 such that $V_s - V_0$ was 4 kV was also determined.

Before each measurement in a run, both V_0 and V_1 were checked by potentiometer. V_s was determined at the start of a run, several times during the run, and at the end of the run. In calculating the data, it was assumed that if V_s increased, it increased linearly with the number of sparks passed.

Correction for the losses in V_1 due to the drop in VT_4 and to capacity division was made by assuming that for a given V_0, the lowest value of V_1 which gives consistent failures corresponds to $\Delta V = 0$. The value of V' thus obtained agrees to within a few percent with the losses as calculated from the known steady-state voltage drop in VT_4 and the measured capacities of the circuit.

3. Experimental results

All time lags reported represent the average of the individual measurements of a run. The minimum time lag in a run was usually about half of the average time lag, while the maximum time lag in a run was about twice the average time lag. The spread in the time lags in a run increases approximately linearly with the average time lag of the run. *Figure 3* shows a typical distribution for four runs at various percent o.v. When the average time lag is of the order of tens of microseconds, the spread is as large as the time lag itself. However, in not a single case was an abnormally short time lag observed. That is, no single measurement gave a time lag which was less than about 40 percent of the average time lag. Thus the measurements represent formative time lags. The fluctuations of the formative time lags cannot be explained by the lack of initiating primary electrons. A part, but not all, of the fluctuations are due to the electronic errors. The other part of the fluctuation is probably inherent in the statistical nature of the spark.

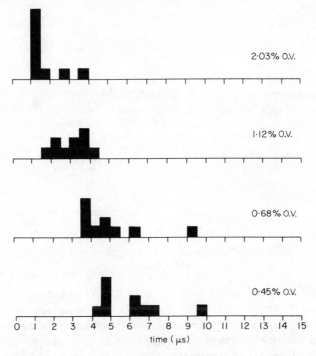

Figure 3. Typical time lag distribution for four runs. $p = 532$
mm Hg, $d = 1$ cm. One square block represents one measurement.

The time lags as a function of percent o.v. for $d = 1$ cm are plotted in *Figure 4*
for four values of the pressure. The term 'low illumination' signifies that the ultra-
violet light used to illuminate the cathode was filtered through several thicknesses of
copper screen to reduce its intensity. The primary current obtained when the
screens were used varied from about 1·3 to about 6 electrons/μs. The term '2000-V
pulse' signifies that V_0 was set at exactly 2 kV below V_s. The principal feature of
Figure 4 is that for all values of the pressures studied, the time lags for very low
percent o.v. are quite long and indeed may increase without limit as the percent
o.v. approaches zero. As the percent o.v. is increased, the average time lags decrease
extremely rapidly, and at about two percent o.v., the time lags are of the order of one
microsecond, in agreement with the results of previous investigators. The measurements below two percent o.v. represent essentially a previously unexplored region.

An additional feature of the curve is the lack of dependence of the time lags on
pressure to within the experimental accuracy of the apparatus.[15] For the four
values of pressure for which the lags are plotted in *Figure 4*, the time lags for a
given percent o.v. are approximately the same. If pressures below several hundred
millimetres Hg were included in the graph, the time lags would follow the curve of
Figure 4 until the percent o.v. reaches a few tenths of one percent. As the percent
o.v. is decreased below this value, the time lags do not increase as rapidly as the

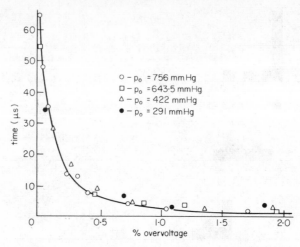

Figure 4. Time lag *vs.* percent o.v., *d* = 1 cm. Low illumination, 2000 V pulse.

curve in *Figure 4*. The departure from the curve at small percent o.v. becomes more pronounced as the pressure is lowered. These shorter time lags for low pressures at overvoltages of a fraction of a percent may be instrumental. (There is an error in V_1 of several volts and for low pressures this error becomes an appreciable percentage of V_s, while for high pressures this error is only a few hundredths of a percent of V_s.)

Figure 5 contains the data of *Figure 4* with the addition of data for *d* = 1·4, 0·6, and 0·3 cm. Here again, there is no dependence of the time lags on pressure. It is seen, however, that the average time lags for a given percent o.v. increase with increasing gap separation. *Figure 6* shows three sections of *Figure 5* for 0·05, 0·1, and 0·2 percent o.v. It is seen that for a given percent o.v., the time lags increase linearly with increasing plate separation.

To determine the possible effects of the illumination and of V_0 on the time lags, additional data were taken at *d* = 1 cm and at various pressures. First, the copper screens were removed from the path of the ultraviolet light. The primary current then ranged from 18 to 20 electrons/μs. The time lag curve is identical with that shown in *Figure 4* to within the experimental error. It can thus be stated that for an average primary current of between 3 and 20 electrons/μs, the time lags are independent of the illumination. Second, the effect of V_0 was investigated at low illumination by changing V_0 from 2 to 4 kV below breakdown, and again to within the experimental accuracy, the data are identical with those shown in *Figure 4*.

To determine the reproducibility of the measurements, duplicate data were obtained at various pressures for *d* = 1 cm (low illumination , $V_s - V_0$ = 2 kV) after the entire experiment was completed. The resulting data again coincide with the results shown in *Figure 4*.

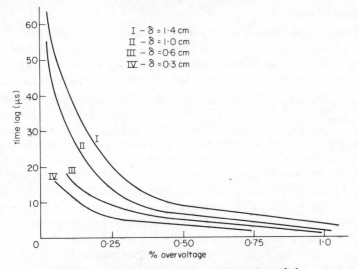

Figure 5. Time lag *vs.* percent o.v. for four values of plate
separation. Curves represent average of data taken at all
pressures between atmospheric and about 200 mm Hg. Low
illumination, 2000 V pulse.

There is an error introduced because of the electronic apparatus, such as the
ripple in V_0 and V_1, the uncertainty in the magnitude of V_1, and the rise time and
decay of V_1. An upper limit of 5 V can be set for the cumulative error of all these
factors. Since this error is essentially independent of V_0, the accuracy of the data
depends on the pressure. Thus at atmospheric pressure and $d = 1$ cm, the lower limit
of quantitatively reliable data is 0·017 percent o.v., while for a pressure of 20 mm
Hg, this lower limit is about 0·3 percent o.v.

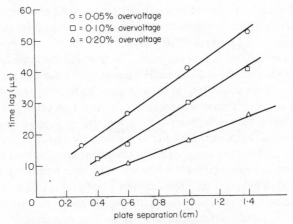

Figure 6. Time lag *vs.* plate separation for three values of
percent o.v. Curves represent average of data taken at all
pressures between atmospheric and about 200 mm Hg. Low
illumination, 2000 V pulse.

4. Discussion of results

The existence of long time lags at low percent o.v. indicates that (for the experimental conditions studied) the motion of positive ions plays an important role in the breakdown process. The fact that the time lags decrease continuously and smoothly as the percent o.v. is increased indicates that the role of the positive ions gradually decreases in importance as the field strength increases, until finally the time lags become so short that the positive ions may be considered to remain stationary throughout the formative time of breakdown.

One can try to account for the results on the basis of the Townsend discharge mechanism, in which secondary electrons are liberated from the cathode by positive ion bombardment. However, for the Townsend mechanism, even at very high overvoltages the time lags should still be in the microsecond region. Since the time lags decrease much more rapidly with increasing percent o.v. than can be explained by the variation of the positive ion velocity, the role of the positive ions is not the emission of secondary electrons from the cathode.

The second possibility for explaining the role of the positive ions in producing the spark is based on the fact that the positive ions aid the spark formation by producing a distorted field in the gap.[16] Electrons produced in the cathode region subsequent to the first avalanche multiply more intensely than in the undistorted field provided α/p increases more rapidly than linearly with E/p, the ratio of field strength to pressure. These newly created electrons may be produced by a secondary mechanism (the most probable ones being photo-ionization in the gas and photoelectric emission at the cathode) or by the constant primary electron current produced by illumination of the cathode. As the positive ions, originally created in the anode region, move toward the cathode, the final strength in the cathode region increases more and more rapidly with time. Thus, owing to the curvature of the α/p *vs.* E/p curve, the quantity $\exp\left(\int_0^d \alpha dx\right)$, which represents the number of ion pairs produced by an electron originating near the cathode, increases with time. If conditions in the gap are not satisfactory for the formation of a spark when the first primary electron has crossed the gap, it is possible that as the positive ions created in the first avalanche approach the cathode, subsequent avalanches may produce the charge densities necessary for the development of a streamer. With such a mechanism, times corresponding to many transit times of a positive ion may occur before the spark develops. As the percent o.v. is increased, the enhanced field necessary for the production of the spark requires a shorter distance of travel of the positive ions and so the time lag decreases. As the percent o.v. is increased further, no motion of the positive ions is required to produce a spark, and a streamer emanates from the anode. With even higher overvoltages, adequate distortion fields for the production of a streamer can be built up even before the primary electron crosses the entire gap. Thus, the distortion field provides a smooth transition from formative times long compared to the positive ion transit time to times comparable to the electron transit time.

The method whereby sufficient initiating electrons are supplied to maintain the pre-spark current is now considered further. It can be shown that if, under the assumption of a primary current with no secondary mechanism, the field strength in the cathode region as a function of time is derived for times less than the transit time of a positive ion (assuming no diffusion and a parabolic dependence of α/p on E/p), and if one assumes that a spark occurs when this distortion field reaches a critical value, fair agreement with the experimental data is obtained. However, if such a mechanism is valid, it is necessary to assume that all the primary electrons are emitted from a small region on the cathode. This is unlikely, since the spark rarely occurred at the same cathode spot twice in succession. Furthermore, no dependence of the time lag on illumination was observed, whereas such a mechanism would depend strongly on the magnitude of i_0. Better agreement with the experimental data is obtained by assuming a secondary mechanism involving photoemission at the cathode by photons produced in the avalanche.[17]

It is believed, therefore, that two processes acting together produce the observed long time lags at low overvoltages. These are (1) a suitable secondary mechanism (most probably photoemission at the cathode) to maintain and perhaps increase the pre-spark current and (2) space-charge distortion due to the large number of positive ions in the gap. It should be noted that such a space-charge mechanism can explain the long time lags obtained for low as well as high pressures. Indeed, in view of the fact that even at the lowest pressures studied (several centimetres of Hg) the time lags for overvoltages of a few percent are as short as those obtained for the higher pressures, it is clear that the classical Townsend theory, hitherto assumed valid for the product pd less than several hundred millimetres Hg, is inadequate.

The search for the transition from the streamer to the Townsend mechanism of spark breakdown has revealed no simple dichotomy in mechanisms as envisaged by Loeb and Meek, and by Raether; the mechanism of the spark depends much more strongly on the percent o.v. than on the value of p and d. These experiments are being continued with other gases.[18]

The authors thank Dr. H. J. White of the Research Corporation and Mr. M. Menes for their aid in developing the electronic circuit, and Professor L. B. Loeb of the University of California for his interest in the problem and for the loan of the chamber used in the research.

References and Notes

1. Loeb, L. B., *Fundamental Processes of Electrical Discharge Through Gases*, John Wiley and Sons, Inc., New York, (1939).
2. Meek, J. M., *Phys. Rev.* **57**, 722, (1940).
3. Loeb, L. B. and Meek, J. M., *The Mechanism of the Electric Spark,* Stanford University Press, Stanford University, California, (1941).
4. Raether, H., *Arch. Elektrotech.* **34**, 49, (1940); *Z. Physik* **117**, 375, 524, (1941); *Ergeb. exakt. Naturw.* **22**, 73, (1949).
5. Unless otherwise noted, all statements in this paper refer to air.
6. Raether takes the value of this product as 1000 mm x cm, see reference 4.

7. Fisher, L. H., *Phys. Rev.* **69**, 530, (1946); Loeb, L. B., *Proc. Phys. Soc.* **60**, 561, (1948).

8. For a detailed description of the background leading to this research, see Fisher, L. H., *Elec. Eng.* **69**, 613, (1950).

9. A somewhat parallel experimental program was undertaken by Gänger, B., *Arch. Elektrotech.* **39**, 508, (1949).

10. Schade, R., *Z. Physik* **104**, 487, (1937).

11. See for example Pedersen, P. O., *Ann. Physik* **71**, 317, (1923); Rogowski, W., *Arch. Elektrotech.* **20**, 99, (1928); White, H. J., *Phys. Rev.* **49**, 507, (1936); Wilson, R. R., *Phys. Rev.* **50**, 1082, (1936); Newman, M., *Phys. Rev.* **52**, 652, (1937); Fletcher, R. C., *Phys. Rev.* **76**, 1501, (1949). The results of these and other authors have been summarized in reference 1, p. 441 and by Strigel, R., *Elektrische Stossfestigkeit* (Verlag. Julius Springer, Berlin, 1939).

12. Sanders, F. H., *Phys. Rev.* **41**, 667, (1932).

13. Posin, D. Q., *Phys. Rev.* **50**, 650, (1936).

14. Fisher, L. H., *Phys. Rev.* **72**, 423, (1947).

15. It is interesting to note that Gänger (reference 9) who studied formative time lags in air in the range from about four to several hundred percent o.v., also found no marked pressure dependence over the range of one atmosphere to 25 mm of Hg.

16. For a discussion of the effect of space charge falsification of the second Townsend coefficient, see Varney, White, Loeb, and Posin, *Phys. Rev.* **48**, 818, (1935).

17. *Editor's note:* Further comments on the calculations referred to by the authors appeared in later papers by Fisher and his colleagues. For example, Kachickas, G. A. and Fisher, L. H., *Pnys. Rev.* **88**, 878, (1952) reported measurements and calculations for nitrogen and concluded that a secondary emission of photoelectrons from the cathode, or from the gas near the cathode, was necessary to explain their observations. A large number of successive electron crossings of the discharge gap was also required. In a later paper by Kachickas, G. A. and Fisher, L. H., *Phys. Rev.* **91**, 775, (1953) on work in argon, the authors concluded that their results 'indicate the universality of the Townsend build-up before breakdown, but the build-up may proceed by different mechanisms in different gases'.

18. Kachickas, G. A. and Fisher, L. H., *Phys. Rev.* **79**, 232, (1950).

Introduction to Papers 8, 9 and 10 (Davidson 'The growth of ionization currents in a uniform field $E(>E_s)$'; Davidson, 'Growth of current between parallel plates'; Davidson, 'Theory of the temporal growth of ionization in gases involving the action of metastable atoms and trapped radiation')

These three short papers by P. M. Davidson are taken from a series of important papers written between 1953 and 1962, and consider the growth of ionisation in a parallel-plate discharge in the absence of space-charge distortion of the applied uniform electric field. Paper 8 was the first of the series and appeared as a mathematical appendix to a paper by Dutton, Haydon and Llewellyn Jones which attempted to account for measurements of formative time lags (such as those described in paper 7 by Fisher and Bederson) on the basis of the same primary and secondary ionisation processes found capable of accounting successfully for the spatial growth of pre-breakdown currents. The secondary processes considered in papers 8 and 9 were the release of electrons from the cathode by positive ions or photons produced in the gas. In paper 10 the action of metastable atoms and of photons which have been repeatedly absorbed and re-emitted on their way to the cathode was considered. A later paper (*Proc. Phys. Soc.* **80**, 143, (1962)) gave further attention to the action of delayed photons.

The series of papers has been usefully summarised by C. Grey Morgan in *Fundamentals of Electrical Discharges in Gases* (Pergamon, (1965)). The treatment given there emphasises those parts of Davidson's analysis which are of particular interest experimentally. H. Raether in *Electron Avalanches and Breakdown in Gases*, Butterworths, (1964) also devotes a chapter (Chapter 7) to the theory of a transient discharge in a uniform electric field. The theory given there is largely based on papers by P. L. Auer (e.g. *Phys. Rev.* **3**, 671, (1958)).

Further Reading

Grey Morgan, C., *Fundamentals of Electrical Discharges in Gases*, Vol. 2, Part 1, of *Handbook of Vacuum Physics*, ed. Beck, Pergamon Press, (1965).
Raether, H., *Electron Avalanches and Breakdown in Gases*, Butterworths, (1964).
Davidson, P. M., *Proc. Phys. Soc.* **80**, 143, (1962).

8

The growth of ionization currents in a uniform field $E(>E_s)$

P. M. DAVIDSON*

Two infinite parallel plates are considered at a separation d, the cathode (at $x = 0$) being exposed to a constant external illumination. At time zero the potential difference of the plates is suddenly increased, and then maintained constant. It is desired to calculate the conditions at any later time (so long as the effects of space charge remain unimportant).

An approximate solution of this problem has already been given by Bartholomeyczyk,[1] who, from a consideration of the differential equations

$$(\partial I_-/\partial t)/v_- = - \partial I_-/\partial x + \alpha I_- \tag{1a}$$

$$(\partial I_+/\partial t)/v_+ = \partial I_+/\partial x + \alpha I_- \tag{1b}$$

and boundary conditions

$$I_-(o, t) = \gamma I_+(o, t) + \delta \int_0^d I_- \exp(-\mu x)\, dx \tag{2a}$$

$$I_+(d, t) = 0 \tag{2b}$$

employs, for calculating current amplification, an expression

$$I_-(x, t) \exp(-\alpha x) = C \exp[\lambda(t - x/v_-)] \tag{3}$$

in which C is a constant and λ is given a real value satisfying $F(d) = 0$, where

$$F(x) = 1 - (\gamma\alpha/\phi)\,[\exp(\phi x) - 1] - (\delta/\psi)\,[\exp(\psi x) - 1] \tag{4}$$

$$\phi = \alpha - (\lambda/v),\ \psi = \alpha - \mu - (\lambda/v_-)$$

and

$$(1/v) = (1/v_-) + (1/v_+)$$

Bartholomeyczyk's expression (3) for I_-, together with an accompanying expression for I_+, satisfy the differential equations (1) and the boundary conditions (2), and are of fundamental importance in the problem. They are not, however, an accurate solution of it, since they do not reduce at $t = 0$ to the charge distributions

* Reproduced from *British Journal of Applied Physics*, 4, 173, (1953) by permission of The Institute of Physics and The Physical Society, London and of the Executor of the late P. M. Davidson.

actually present at that time; moreover, the boundary condition (2a) omits an additive constant, say I_0, due to external illumination. (Gugelburg[2] has noted one of these defects—the omission of the constant in the boundary condition—but has not proposed a valid way of correcting for its effects.) It is thus difficult to estimate to what extent the simple expression (3) will approximate to the true solution, or what value should be assigned to the constant C in order to give the best agreement with the true solution.

Before proceeding to the exact solution, it should be noted that Bartholo-meyczyk's formula can be altered so as to satisfy equations (1) and the corrected boundary conditions (but still not the correct initial conditions) by the following procedure. If λ is given Bartholomeyczk's value, and, if

$$P = 1 - \gamma \left[\exp (\alpha d) - 1\right] - [\delta/(\alpha - \mu)] \{\exp [(\alpha - \mu)d] - 1\}$$

then a short calculation shows that an expression

$$I_-(x, t) \exp (-\alpha x) = (I_0/P) + C \exp [\lambda(t - x/v_-)] \tag{5}$$

with an accompanying expression for I_+ satisfy equations (1) and the corrected boundary conditions. The constant C may be given a value which gives rough agreement with the initial conditions; for example, if the initial value of $I_-(o, t)$ is c, the solution can be made to have that initial value of $I_-(o, t)$ by taking C as $c - I_0/P$.

To see how much inaccuracy still remains, a solution must be found which not only satisfies equations (1) and the correct boundary conditions, but also satisfies the initial conditions completely, that is, gives a prescribed initial distribution of positive and negative charge in the whole range $x = 0$ to d. The nature of this exact solution may be simply stated. The equation $F(d) = 0$ has, in addition to the real root, an infinite number of complex roots; the last term in equation (5) must be replaced by a summation containing these various λ's and with C's determined by the initial conditions. The mathematical problem is of little interest, and it will be sufficient to state the final expressions obtained for I_- and I_+ (the latter being of importance for estimating when the effects of space charge cease to be unimportant).

It is convenient to think of the expression for, say, $I_-(x, t)$ as consisting of two parts: (i) the value which it would have if the initial charge distribution had been absent; and (ii) the value which it would have if the constant generation I_0 had been absent. The quantity $I_-(x, t)$ which will be present in the experiment is the sum of the two parts. The same remarks apply to $I_+(x, t)$.

In the part (ii), $I_-(x, t)$ and $I_+(x, t)$ are conveniently calculated from the initial distribution, say $\rho_- = f_-(x)$ and $\rho_+ = f_+(x)$, by means of four g's (Green's functions), of which, for example, $g_-^+(xtx_1)$ is the $I_-(x, t)$ due to the presence at time zero of unit positive quantity in the region of $x = x_1$. Thus the required $I_-(x, t)$ of part (ii) is

$$\int_0^d [g_-^- f_-(x_1) + g_-^+ f_+(x_1)] \, dx_1$$

and a corresponding expression gives $I_+(x, t)$.

At t's less than x/v_- the values of g_-^- and g_+^-, and also of the I_- and I_+ of part (i), are of a simple nature, readily seen on visualizing the motion. [For example, g_-^- is zero in this range of time if $x < x_1$. If $x > x_1$ there is a sudden pulse at $t = (x - x_1)/v_-$, conveying a quantity $\exp \alpha(x - x_1)$.] At all later times the series expressions given below for these quantities are valid. Similarly the values of g_-^+ and g_+^+ are readily seen at times up to $(x_1/v_+) + (x/v_-)$ after which the series are to be used.

The series expressions are:

Part (i).

$$I_-/I_0 = \exp (\alpha x)/P + \Sigma Q$$
$$I_+/I_0 = [\exp (\alpha d) - \exp (\alpha x)] /P + \Sigma RQ$$

where $Q = (\lambda D)^{-1} \exp \{[\alpha - (\lambda/v_-)] x + \lambda t\}$

$R = (\alpha/\phi) \{\exp [\phi(d - x)] - 1\}$

$D = (\gamma \alpha/\phi^2 v)[1 - (1 - \phi d) \exp (\phi d)] + (\delta/\psi^2 v_-)[1 - (1 - \psi d) \exp (\psi d)]$

Part (ii).

$$g_-^- = \Sigma G_-^- \qquad\qquad g_+^- = \Sigma R G_-^-$$
$$g_-^+ = \Sigma G_-^+ \qquad\qquad g_+^+ = \Sigma R G_-^+$$
$$G_-^- = \lambda Q F(x_1) \exp \{[(\lambda/v_-) - \alpha] x_1\}$$
$$G_-^+ = \gamma \lambda Q \exp [- \lambda x_1/v_+]$$

References
1. Bartholomeyczyck, W., *Z. Phys.* **116**, 235, (1940).
2. Von Gugelburgh, H. L., *Helv. Phys. Acta* **20**, 250 and 307, (1947).

9

Growth of current between parallel plates

P. M. DAVIDSON*

For assessing the relative magnitude of the various regenerative processes which can occur in a gaseous discharge, considerable importance attaches to the theory of current growth. Of particular interest is the growth of current, in absence of space-charge distortion of the field, between parallel plates maintained at a constant potential difference, and with a constant electron generation I_0 maintained at the cathode by external illumination, and with secondary electron generation due to the action of photons and positive ions on the cathode. The writer has given[1] the exact solution of this problem, and also an approximate solution. Bartholomeyczyk had given an approximate solution which neglected the maintained I_0, and was thus inapplicable to the present problem, except for tracing, very roughly, the current growth in case of high overvoltage. In the writer's approximate solution (which does take account of the maintained I_0) the electron current $I_-(0, t)$ at $x = 0$ (the cathode) at time t is of the form

$$A - B \exp(\lambda t),\tag{1}$$

and the constant B is not specified exactly.

A recent paper by Bandel,[2] describing his experimental work on current growth and its interpretation, is of considerable interest, but his mathematical discussion of current growth in absence of space-charge distortion is inaccurate.[3] He considers the case in which, at time zero, there are no charged particles between the plates [the case called Part (i) in my exact solution]. He regards my exact solution as impracticably cumbersome; on the other hand he maintains that at times greater than d/v (or d/v_- if γ is zero[4]) my approximate solution, with a correct choice of the constant B is the exact solution of the problem. This contention is not correct, and the mathematical argument by which it is obtained is erroneous, for the following reason.

* Reproduced from *The Physical Review*, **99**, 1072, (1955) by permission of The American Physical Society, New York and of the Executor of the late P. M. Davidson.

The integral equation (5) in Bandel's paper specifies the desired solution throughout all (positive) times. My approximate expression (1), with any value of B, satisfies the integral equation at t's greater than a specified value; but it does not follow that, with a correctly chosen B, it is the desired solution in that range of t. It can, in fact, be shown that the differential equations and boundary conditions [equations (1) and (2) in Bandel's paper] require that if the expression (1) holds exactly after a finite time it must have held exactly at all (positive) times. Moreover $I_-(x, 0)$ must have been, not zero as the problem requires, but $\exp(\alpha x)\{A - B \exp(-\lambda x/v_-)\}$. Except in the case when γ is zero, $I_+(x, 0)$ must also have been a nonzero distribution. Such facts are hardly surprising, for, as was pointed out in my paper, the equation for λ has, in addition to its obvious root which appears in (1) an infinite number of others, which appear in my exact solution. It would be surprising if the desired solution, in a certain range of time, could be constructed exactly by using only the one root.

In support of his contention, Bandel gives expressions which he says are the exact solution in the case of δ zero. It may readily be verified, however, that this is not so; there is a range of t (between d/v and $2d/v$) in which his expressions fail, and the alteration which they need in that range leads to alterations in all later ranges. This may be expressed explicitly by writing my exact solution in an alternative form. Although for most purposes it is, as we shall see, less convenient than the original form, it demonstrates the present point. For δ zero, $I_-(0, t)$ in the range nd/v to $(n + 1)d/v$ is

$$I_0 \sum_{m=0}^{n} \gamma^m \exp(m\alpha d)\left\{ (1+\gamma)^{-(m+1)} + \frac{(-\alpha v)^m}{m!} \right.$$

$$\left. \times \left[\frac{d^m}{dp^m}\left\{ \left(1 - \frac{\alpha v}{p}\right) \exp[p(t - md/v)]\right\}\right]_{p = p_0} \right\}, \quad (2)$$

where $p_0 = \alpha v(1 + \gamma)$ and $0!$ means unity. Thus when $t < d/v$, it is $I_0[(1 + \gamma \exp(p_0 t)]/(1 + \gamma)$, and when $d/v < t < 2d/v$, it is

$$\frac{I_0}{1+\gamma}\left\{ \left(1 + \frac{\gamma \exp(p_0 t)}{1+\gamma}\right) \right.$$

$$\left. + \gamma \exp(p_0 t) \left(1 - \frac{\{1 + \gamma p_0(t - d/v)\}\exp(-\gamma\alpha d)}{1+\gamma}\right)\right\}$$

and so on (extra terms being added in each successive range). Comparing this with Bandel's expression, we see that the latter is correct only up to $t = d/v$; as soon as it becomes identical with my approximate expression it ceases to be exact. My approximate expression does not become exact after any finite time. For one thing, it is too mathematically smooth; in the exact expressions, given above, only the first n time differentials are continuous at the point $t = (n + 1)d/v$.

If γ and μ are zero, but not δ, the exact expressions above remain valid on changing γ to δ/α, and v to v_-.

If neither γ nor δ is zero, but only μ, the exact expression, in the form corresponding to (1), contains two p's, say p_1 and p_2, the roots of

$$1 + \frac{\delta}{\psi} + \frac{\alpha\gamma}{\phi} = 0,$$

where $\psi = \alpha - (p/v_-)$ and $\phi = \alpha - (p/v)$. When t is less than d/v and is in the range from nd/v_- to $(n+1)d/v_-$ the expression for $I_-(0, t)$ is

$$I_0 \sum_{m=0}^{n} \left\{ \left(\frac{\delta}{\alpha}\right)^m \exp\,(m\alpha d) \left(1 + \gamma + \frac{\delta}{\alpha}\right)^{-(m+1)} \right.$$

$$\left. + \sum_i \frac{\delta^m (vv_-)^{m+1}}{m!} \left[\frac{d^m}{dp^m}\, \frac{\psi\phi^{m+1} \exp\,(pt + m\psi d)}{p(p - p_2)^{m+1}} \right]_{p=p_1} \right\} \quad (3)$$

where Σ_i means a summation over an interchange of the suffixes 1 and 2. Thus, when t is less than d/v_-, it is

$$I_0 \left\{ \left(1 + \gamma + \frac{\delta}{\alpha}\right)^{-1} + \frac{vv_-}{p_1 - p_2} \left(\frac{\psi_1\phi_1 \exp\,(p_1 t)}{p_1} - \frac{\psi_2\phi_2 \exp\,(p_2 t)}{p_2} \right) \right\},$$

and so forth.

From such expressions as (2) and (3) the values of $I_-(x, t)$ and $I_+(x, t)$ are given by the formulas

$$I_-(x, t) = \exp\,(\alpha x) I_- \left(0, t - \frac{x}{v_-}\right);$$

$$I_+(x, t) = \int_x^d \alpha I_- \left(0, t + \frac{x}{v_+} - \frac{x'}{v}\right) \exp\,(\alpha x')dx',$$

in which $I_-(0, T)$ is to be set equal to zero at T negative.

There are several ways of deriving the exact expressions, either in the form such as (2) and (3) or in the form given in my previous paper. If integration in the complex plane is employed the calculation of $I_-(0, t)$ depends on evaluating $\int C \{\exp\,(pt)f(p)/p\}\, dp$ where C is an infinite semicircle on the right of the imaginary axis, and $f(p)$ is

$$\left\{ \left(1 + \frac{\delta}{\psi} + \frac{\alpha\gamma}{\phi}\right) - \left(\frac{\delta}{\psi}\exp\,(\psi d) + \frac{\alpha\gamma}{\phi}\exp\,(\phi d)\right) \right\}^{-1},$$

which has an infinite number of poles. Integrating in the usual way yields the expression for $I_-(0, t)$ given in the section called Part 1 in my previous paper. The alternative form given in the present paper may be obtained by making a binomial expansion of $f(p)$ on the contour C and deleting all terms which, at a given t, will (owing to their character at infinity) give zero integrals. We thus have to integrate a finite number of terms, each having only a few poles. Expressions such as (2) and (3) are obtained.

The expressions such as (2) and (3) are chiefly of interest in the case of a discharge which is highly overvolted, so that even during the first few n's the current amplification becomes very great, and thus of experimental interest. These exact expressions, containing only a few terms, are of value in such cases. But whatever the voltage the form given in my previous paper is more convenient throughout the subsequent growth. That form specifies a definite value of B, say B_0, in the approximate formula, and the correction terms which must be added to make the expression exact. A few remarks concerning it may be useful to experimentalists. To visualize the nature of B we may write $A - B \exp(\lambda t)$ as $(I_0/P)\{1 - \exp[\lambda(t + \Delta)]\}$. Consider, for example, the case in which γ and μ are zero. Setting Δ zero makes this approximate expression for $I_-(0, t)$ smaller than the exact value at all positive t's (as may be seen from the nature of the error in the initial distribution); on the other hand setting Δ equal to d/v_- makes the expression too large at all positive t's. If the magnitudes of A and αd are large compared with unity, B_0 makes Δ about $d/2v_-$; (the B which Bandel assumes makes Δ nearly d/v_-). In using the exact formula to estimate whether, at a given t, the correction terms are of any importance, there is no need to insert exact values of the complex λ's. It would be sufficient, in this case of γ and μ zero, to write $\lambda_{n\pm} = \bar{\lambda}_n \pm (2\pi i n v_-/d)$ where $\bar{\lambda}_n = \lambda_0 - (2\pi^2 n^2 v_-/\alpha^2 d^3)$ at small n's and tends to $-(v_-/d)\log(2\pi n/\alpha d)$ at very large n's. λ_0 is the real λ, which appears in the B_0 term. Thus the correction terms in $I_-(x, t)/I_0$ are approximately

$$\sum_{n=1}^{\infty} \frac{1}{\pi n} \exp\{\bar{\lambda}_n[t - (x/v_-)]\} \cos\left\{\frac{2\pi n v_-}{d}\left(t - \frac{x}{v_-}\right) - \theta_n\right\},$$

θ_n being the angle, less than π, whose tangent is $2\pi n v_-/\bar{\lambda}_n d$.

In using the accompanying formula[1] for $I_+(x, t)$, it is convenient to hasten the convergence at the smaller values of t by subtracting a series which is zero; for example, so long as $v\{t + (x/v_+)\}$ is less than d we may evidently subtract $I_+[v\{t + (x/v_+)\}, v\{t + (x/v_+)\}/v_-]$ that is, we may replace the d's in the numerators of the expression for $I_+(x, t)$ by $v\{t + (x/v_+)\}$. If the formula is used without a procedure of this type, the case of γ zero must be regarded as the limiting case of γ very small.

If the distributions of electrons and ions between the plates at time zero are not negligible, the expressions called Part (ii) in my previous paper are required.

Note added in proof.—For the case in which there is only one regenerative process, an expression equivalent to the alternative form of the exact solution has been obtained in a recent paper by P. L. Auer [*Phys. Rev.* 98, 320, (1955)]. His expression (3.17), which he proposes for practical use, is an approximate formula for $I_-(0, t)$ at integral values of n (with $n = 1$ at $t = 0$) for the case of γ zero and $\delta d \ll 1$; but he underestimates the error in this expression, at large n's in the case of $V > V_s$. The expression differs from my approximate formula $A - B_0 \exp(\lambda t)$ in two ways. Firstly, λ has been replaced by a different quantity; it is practically the value, say λ_∞, which λ assumes if αd is increased to infinity while A is kept unchanged; and secondly, B_0 has been changed to a value which makes the expression practically correct at time zero. It may be shown that λ_∞/λ is approximately $\alpha d/(\alpha d + 1)$ if αd is fairly large compared with unity (values about ten are common in the experimental work). Thus, if $V > V_s$ his alteration in λ makes his expression (3.17) too small at large n's, the true current exceeding it by a percentage which increases without limit at very large n's. It would have been better to have retained my approximate formula $A - B_0 \exp(\lambda t)$, in which, of course, the percentage error diminishes to zero at large n's, when the omitted oscillatory terms (the terms with complex λ's) have died out relative to the retained terms.

The physical reason for the accuracy of Auer's formula at small integral n's may be seen by imagining αd increased to a very large value without changing A. Then, owing to the shape of the curve $\exp(\alpha x)$, an avalanche generated by an electron at the cathode produces its photoelectrons very suddenly, just before it reaches the anode. Thus the $I_-(0, t)$ curve due to a maintained I_0 becomes practically a staircase during the first few n's, the sudden steep slopes occurring just before the integral n's. If now we diminish αd to a more practical value, we diminish thereby the mean time at which an avalanche, after setting out from the cathode, generates its photoelectrons; we diminish it from d/v_- to about $(\lambda_\infty/\lambda)(d/v_-)$. Thus we may say that we now have a more rapid amplification process, though the effect will not appear appreciably at the first few integral n's. In that range the extra current will appear at intermediate points, as a less steep ascent in the regions approaching the integral n's; but at later times the effect will extend to the integral n's. It should be noted, moreover, that in the range of very small n's a smooth curve drawn through the points given by Auer's formula is only correct at the integral n's, and usually one is not especially interested in those particular points; if a smoothed curve of current growth is required it would have been better to have retained my approximate formula $A - B_0 \exp(\lambda t)$ even in this range of small n's. For example, in the extreme case of $\alpha d \gg 1$, which makes the true curve almost a staircase in this range, Auer's smooth curve is like an inclined (and rather curved) rod resting on the stairs; its error never changes sign. The curve drawn from my approximate formula intersects each stair, giving an error alternately positive and negative, and is a good smoothed approximation.

References and Notes

1. Dutton, Haydon, and Jones, *Brit. J. Appl. Phys.* **4**, 170, (1953), Mathematical Appendix by P. M. Davidson.
2. Bandel, H. W., *Phys. Rev.* **95**, 1117, (1954).
3. For his purposes, the resulting error is not at all serious in his ultimate numerical applications, which are confined to the case in which positive ions do not generate electrons at the cathode.
4. This notation is that of my previous paper (reference 1).

10

Theory of the temporal growth of ionization in gases, involving the action of metastable atoms and trapped radiation

P. M. DAVIDSON*

1. Introduction

In recent years much experimental work has been carried out on the temporal growth of ionization currents in gases in uniform electric fields, especially in the case when the applied field E exceeds the value E corresponding to breakdown potential. The rate of increase of current depends greatly on the nature of the secondary ionization processes, and extremely rapid rates of growth are obtainable, for example, in hydrogen, where the predominant secondary processes are those of cathode emission due to the incidence of positive ions and unscattered photons originating in the gas. An account of this experimental work is given in the monograph by Llewellyn Jones (1957). The present writer (Davidson, 1953) has given exact and approximate formulae for the current growth which have enabled the experimental data to be analyzed when these two processes are acting simultaneously. The rate of amplification naturally depends on the degree of overvoltage; with high overvoltage ($\approx 100\%$) only a few gap transit times for electrons may be sufficient to produce great amplification, whereas with low overvoltage ($< 5\%$) many transit times are required to produce the same amplification. The writer has therefore given his formulae in a variety of algebraical forms (Davidson, 1955, 1956, 1957), convenient for use in these different cases.

It has been pointed out by Professor Llewellyn Jones that the author's formulae ought, for completeness, to be extended so as to take account of a type of secondary action quite different from those considered above. It will be noted that in those processes the method by which the active particles (photons or positive ions) move to the cathode is not one of diffusion. The ions move to it with a steady velocity in the electric field, and relatively negligible diffusion; the unscattered photons move to it almost instantaneously.

In some circumstances, however, the motion of photons may be regarded as one of diffusion. Photons emitted by atoms which have been excited by the electron

* Reproduced from Proceedings of the Royal Society, **A249,** 237, (1959) by permission of The Council of The Royal Society, London and of the Executor of the late P. M. Davidson.

current may be strongly absorbed by the unexcited atoms and, after a mean time interval τ, re-emitted with scattering. During the period τ they may be thought of as being trapped in the atom. They thus proceed to the electrodes by a process of diffusion, due to repeated scattering. It is usually accompanied by appreciable destruction in the gas.

Further, it is well known that photons and positive ions are not the only active particles which can produce cathode electron emission (Llewellyn Jones, 1957, chap. 1.4.4). It can be produced by the incidence of excited atoms in metastable states. A metastable atom, generated in the gas, may, by the collisions which it makes in the gas, be destroyed (that is, reduced to the ground state), in some cases with generation of an active photon; or it may reach the cathode, in which case it may not only be reduced to the ground state but may thereby liberate an electron. Since the metastable atoms are uncharged their motion to the electrodes is due entirely to diffusion, and the resulting current amplification will usually be much slower than that produced by unscattered photons or by positive ions driven by the electric field. (Approximate calculations for the rate of current growth due to metastable atoms have been made in papers by Engstrom and Huxford (1940); Newton (1948) and Molnar (1951). This work will be discussed in section 4.)

The formulae which we shall employ to represent the diffusion of repeatedly scattered photons resemble the formulae which represent the diffusion of metastable atoms, and space will be saved by taking advantage of this resemblance. The two cases will be referred to as case (b) and case (a), respectively.

To investigate the spatial and temporal growth produced by processes depending on diffusion, it will at first be assumed that the only secondary process operating is of this type. It will further be assumed, both in case (a) and in case (b), that the internal destruction produces no active photons capable of reaching the cathode either directly or by diffusion. Clarity is thereby gained, and there is no difficulty in generalizing the treatment to obtain a perfectly general case, in which all possible secondary processes are acting simultaneously.

2. The continuity equations and boundary conditions

Considering a region in the gas, at distance x from the cathode, we must write the diffusion equation satisfied by the active particles (metastable atoms in case (a) and photons in case (b)). Let $n(x, t)$ be their spatial density, and $j(x, t)$ their current density in the x-direction. In case (b) we include in n (and in j) both the bound and free photons. Thus, writing $i_-(t)$ for the electron current density at the cathode, and W_- for the electron drift velocity, we have

$$\frac{\partial n(x, t)}{\partial t} = -\frac{\partial j(x, t)}{\partial x} + \alpha_1 \exp(\alpha x) i_- \left(t - \frac{x}{W_-}\right) - \frac{n(x, t)}{\tau_1}. \tag{1}$$

Here α is the number of electrons and α_1 the number of active particles which an electron generates in travelling unit distance in the x-direction, and $1/\tau_1$ is the fraction of the active particles in any region which are destroyed per unit time by their collisions with unexcited atoms. We assume that i and n are small enough to enable us to neglect quantities proportional to their product (e.g. the destruction of metastable atoms by electrons). Thus in cases of progressive current growth the treatment ultimately fails, both on this account and also on account of space-charge distortion of the electric field.

To treat the equation we calculate a quantity D, a diffusion coefficient, which makes

$$j = -D \, dn/dx \tag{1a}$$

a correct expression when applied to a steady state, not changing with time; and if this D is constant in space we write the term $-\partial j(x,t)/\partial x$ in the time-varying equation (1) as $D\partial^2 n(x,t)/\partial x^2$, though it is not strictly accurate. Thus (1) becomes

$$\frac{\partial n(x,t)}{\partial t} = D \frac{\partial^2 n(x,t)}{\partial x^2} + \alpha_1 \exp(\alpha x) \, i_-\left(t - \frac{x}{W_-}\right) - \frac{n(x,t)}{\tau_1}. \tag{2}$$

The diffusion coefficient D in case (a), being an ordinary atomic or molecular diffusion coefficient, is of too familiar a nature to need discussion. It may be written approximately as $\frac{1}{3}l\bar{v}$, where l is a suitably defined mean free path of the metastable particles, and $\bar{v}$ their mean kinetic velocity. We may assume that $l/\bar{v}\tau_1$ is a small fraction, since otherwise groups of metastable particles generated in the gas are practically all destroyed after making only a few collisions, and such cases are of no practical interest.

In case (b) we write n_b and n_f for the spatial densities of the bound and free photons. Thus $n = n_b + n_f$. Of the free photons let the fraction which become bound per unit time be c/l, and on re-emission let them retain a fraction F of their original resultant momentum. As in the scattering of atoms, we will call l the collision mean free path and l_1 defined as $l(1 - F)$, the momentum mean free path. By analogy with case (a) we regard $\{\tau + (l/c)\}/\tau_1$ as a small fraction, since otherwise the case is of no practical interest. Then the internal destruction, though contributing an important term n/τ_1 to the equation (2) will have little effect on the coefficient D in the equation. Thus in calculating D we may without much error regard the internal destruction as being absent. As remarked above, D is calculated by considering a steady state; it is one in which the active particles have a pressure gradient, causing their directional distribution to depart somewhat from spherical symmetry, and thus causing them to have a resultant current density. Let the current density of the free photons, each of energy ϵ_0, be j_f. Then since each of them has a momentum of magnitude ϵ_0/c, their resultant momentum, say M, in unit volume, is $(j_f/c)(\epsilon_0/c)$, which is $j_f\epsilon_0/c^2$. By the definition of l_1, the rate of destruction of M in the unit volume by collisions is cM/l_1, which is $j_f\epsilon_0/cl_1$. Owing to the pressure gradient the flow of free photons through the faces of the unit volume is increasing M at a rate given approximately by $-d(\frac{1}{3}\epsilon_0 n_f)/dx$. Thus, since there is a steady state, we have $j_f = -\frac{1}{3}l_1 c(dn_f/dx)$. To write the n_f in this expression in terms of n we recall that τ may be defined as the mean time for which a photon remains bound before being released, or as the mean lifetime of the excited atomic state; or we may say that n_b/τ is the rate at which n_b excited atoms are radiating photons. The three definitions are equivalent. Thus we have $(n_b/\tau) = (cn_f/l)$, expressing the fact that the rate at which photons are becoming bound equals the rate at which they are being released. Thus, since $n = n_b + n_f$, we have $n = n_f\{(cr/l) + 1\}$ and the above expression for j_f becomes

$$j_f = -\frac{\tfrac{1}{3}ll_1}{\tau + (l/c)}\frac{dn}{dx} = -\frac{\tfrac{1}{3}l^2}{\{\tau + (l/c)\}\{1 - F\}}\frac{dn}{dx}.$$

Thus, if we neglect the current density j_b of the bound photons the D in case (b) is given approximately by

$$D = \tfrac{1}{3}l^2/\{\tau + (l/c)\}\{1 - F\}. \tag{3}$$

Taking account of j_b only increases D by an amount which is of the order of an atomic diffusion coefficient, like the D in case (a). If expression (3) is itself only of this order, case (b) is not very different from case (a). If, on the other hand, the approximate expression (3) is the dominant term in D, it is not certain that on inserting it in (2) we shall obtain an accurate equation, with D practically constant in space and time; for it must be remembered that the current of photons is not accurately monochromatic, and thus, since their absorption by atoms is a resonance phenomenon, any spatial variation in their frequency distribution will be accompanied by a spatial variation in l and hence in D. Published theoretical treatments (see Fowler, 1956) are not in agreement as to whether this variation together with the approximations mentioned above, will cause an equation such as (2), with constant D, to be appreciably inaccurate when applied to trapped radiation in the gap. Until this disagreement is resolved, we cannot be certain that equation (2) with constant D applies to trapped radiation with the high accuracy with which it applies to metastable atoms.

We consider now the boundary conditions of equation (2) at the electrodes. Of the number, say N, of active particles which strike the cathode per unit time let a fraction g be destroyed by so doing. If g is not greater than, say $\tfrac{1}{5}$, the resulting asymmetry in the directional distribution of the active particles near the cathode is not sufficient to cause a serious failure of the equations there. Thus if $g < \tfrac{1}{5}$ we may write approximately, at the cathode, in case (a), $N = \tfrac{1}{4}n\bar{v}$ and $D(\partial n/\partial x) = g\tfrac{1}{4}n\bar{v}$, that is

$$n = h\frac{\partial n}{\partial x} \quad \text{at} \quad x = 0, \tag{4}$$

where $$h = 4D/g\bar{v} = \frac{4}{g\bar{v}}\tfrac{1}{3}l\bar{v} = \frac{4}{3}\frac{l}{g}. \tag{5}$$

Similarly, let a fraction G of the active particles which strike the anode be destroyed by so doing. Then

$$n = -H\partial n/\partial x \quad \text{at} \quad x = d, \tag{6}$$

H being $$\tfrac{4}{3}l/G. \tag{5a}$$

Of the active particles destroyed at the cathode let a fraction g_1 cause the emission of an electron. Thus if there is an externally maintained electron current density I_0 there, we have

$$i_- = I_0 + g_1 D(\partial n/\partial x) \quad \text{at} \quad x = 0. \tag{7}$$

Thus at such values of g and G the boundary conditions are (4), (6), (7).

If g and G are larger fractions, say greater than $\frac{1}{3}$, (4) and (6) will, owing to the smallness of l compared with such distances as d, become practically $n = 0$. Strictly speaking, the above derivation does not hold for such large values of g and G, because owing to the great asymmetry in the directional distribution of the active particles near the electrode the expression (1a) and the differential equation will fail there. However, a well-known argument (used in treating the diffusion of charged particles) shows that for these large values of g and G we may regard the diffusion equation and expression (1a) as holding right up to the electrode, and n as becoming zero there (see, for example, Pidduck, 1925).

Turning now to case (b) we have to replace the D in (7) by the expression (3); and the expressions leading to (5) have to be replaced by $N = \frac{1}{4}n_f c = \frac{1}{4}nc/\{(c\tau/l) + 1\}$; thus we have at $x = 0$

$$D(\partial n/\partial x) = \frac{1}{4}gnc/\{(c\tau/l) + 1\}.$$

Inserting the expression (3) for D, we see that in case (b), the h in (4) is $\frac{4}{3}l/(1 - F)g$. Similarly, H is $\frac{4}{3}l/(1 - F) G$. As in case (a), h and H may be given the values zero if g and G are more than small fractions.

3. Solutions for a steady state

Before proceeding to the solutions for temporal growth, we consider the solutions for a steady state (n and i_- constant in time). We then have

$$D\frac{d^2n}{dx^2} + \alpha_1 i_- \exp{(\alpha x)} = \frac{n}{\tau_1}$$

with boundary conditions (4), (6), (7).

The equation can be immediately integrated, giving

$$\frac{i_-}{I_0} = 1 \left/ \left\{ 1 - \frac{\delta_1}{\alpha^2 - \mu^2} \left[\frac{(1 + \alpha H)\,[\exp{(\alpha d)} - \cosh \mu d] - (\alpha + H\mu^2)\,(\sinh \mu d)/\mu}{(H + h)\cosh \mu d + (1 + Hh\mu^2)\,(\sinh \mu d)/\mu} \right] \right\} \right., \tag{8}$$

where $\delta_1 = g_1 \alpha_1$ and $\mu = 1/\sqrt{(D\tau_1)}$. Provided g and G are large enough to make

$$\alpha H, (H + h)/d \quad \text{and} \quad \mu(H + h) \text{ small fractions}, \tag{8a}$$

all terms in (8) containing H and h can be omitted without much error, and it thus reduces to

$$\frac{i_-}{I_0} = 1 \left/ \left\{ 1 - \frac{\delta_1}{\alpha^2 - \mu^2} \left[\frac{\exp{(\alpha d)} - \cosh \mu d - \alpha(\sinh \mu d)/\mu}{(\sinh \mu d)/\mu} \right] \right\} \right., \tag{9}$$

which is the exact solution if $h = H = 0$.

Thus, when μd (but not αd) is a small fraction, equation (9) becomes practically

$$\frac{i_-}{I_0} = 1 \left/ \left\{ 1 - \frac{\delta_1}{\alpha^2 d} \left[\exp(\alpha d) - \alpha d - 1\right] \right\} \right. ,$$

which, if αd is greater than about 5, is practically

$$\frac{i_-}{I_0} = 1 \left/ \left(1 - \frac{\delta_1 \exp(\alpha d)}{\alpha^2 d} \right) \right. . \tag{10}$$

On the other hand, if $\alpha d > 5$ and μd is between, say, unity and $\frac{3}{4}\alpha d$, equation (9) becomes practically

$$\frac{i_-}{I_0} = 1 \left/ \left\{ 1 - \left[\frac{2\delta_1 \mu d \exp(-\mu d)}{(\alpha^2 - \mu^2)d} \right] \exp(\alpha d) \right\} \right. . \tag{11}$$

It must be remembered that i_- is the electron current density at the cathode. The total current density, say I, being the electron current density at the anode, is $i_- \exp(\alpha d)$. Formula (10) is obviously analogous to Townsend's formulae for other secondary processes. As in those cases, a potential V_s may be defined as the applied potential which makes the denominator vanish. If the denominator becomes negative the expression becomes physically meaningless, indicating that no steady state exists.

If G and μ are infinitesimal, but not g (that is, if active particles are destroyed at the cathode, but not in the gas or at the anode), H is infinite, h finite and μ infinitesimal. Thus, (8) becomes

$$\frac{i_-}{I_0} = 1 \left/ \left\{ 1 - \frac{\delta_1}{\alpha} \left[\exp(\alpha d) - 1\right] \right\} \right. . \tag{12}$$

This case is easily visualized, since the entire number $\alpha_1 \left[\exp(\alpha d) - 1\right]/\alpha$ of active particles generated in a single avalanche (due to one electron starting from the cathode) is eventually destroyed at the cathode, liberating $\delta_1 \left[\exp(\alpha d) - 1\right]/\alpha$ electrons; and if this number is only infinitesimally less than unity, i_- will ultimately become infinite, since the generation I_0 is maintained.

Concerning these various formulae the following points may be noted:

(i) Provided the conditions (8a) are satisfied, (8) reduces to (9) independently of the ratio H/h, that is g/G.

(ii) If there is only slight internal destruction, nearly all the active particles generated are destroyed at the electrodes, but only a fraction $(1/\alpha d)$ of them is destroyed at the cathode (provided αH and $(H + h)/d$ are small fractions and $\alpha d > 5$). This is seen by comparing (10) and (12). The reason is that most of the active particles are generated much nearer the anode than the cathode, and thus there is more destruction at the anode than at the cathode.

(iii) Suppose that at $V = V_s$, αd is about 10. Then we see from (10) and (11) that if μd is a small fraction, $\delta_1 d$ is only about exp (-5); but it has a much larger value if μd is, say, $\frac{3}{4}\alpha d$. It may be shown from (9) that if $\mu = \alpha$ at $V = V_s$, then $\delta_1 d$ is about unity at $V = V_s$ (provided αd is greater than about 5 and provided the conditions (8a) are satisfied).

(iv) It is well known that the Townsend formula for the case when there is secondary emission in the gas produced by positive ions, and secondary cathode emission produced by positive ions, directly transmitted photons and metastable atoms, may, subject to certain conditions concerning the magnitudes of the coefficients, be written approximately as

$$I/I_0 = \exp{(\alpha d)}/\{1 - \omega\,[(\exp{(\alpha d)} - 1)]\,/\alpha\},$$

where
$$\omega = \beta + \alpha\gamma + \delta + \epsilon.$$

Here ϵ is a secondary coefficient due to the metastable atoms, and the other symbols have their usual meanings. By extending the calculation (Llewellyn Jones, 1957, p. 55) which leads to this formula it may easily be shown that if the conditions leading to (10) hold, ϵ is $\delta_1/\alpha d$. (If both metastable atoms and trapped radiation are acting, the δ_1 in this expression must be replaced by the sum of the two δ_1's of the separate processes.) It will be noted that at constant E and p, this term ϵ (unlike β, $\alpha\gamma$, and, in absence of absorption, δ) varies with d, owing to the factor $1/\alpha d$, but the variation is slight over the short range in which I/I_0 is very large.

4. Solutions for temporal growth

In considering temporal growth it is useful to begin by noting some orders of magnitude. For the moment we neglect the internal destruction. If an active particle, generated at x, diffuses to the cathode, the average time which it takes to do so is of order x^2/D. Thus, if x is a fairly large fraction of d, this time, say T, is of order $d^2/l\bar{v}$ in case (a). For positive ions the gap transit time, say T_+ is d/W_+ where W_+ is the drift velocity of the ions. Thus, in case (a), T/T_+ is of order $(d/l)\,(W_+/\bar{v})$, and this, owing to the factor d/l will usually be a very large number. In case (b), T/T_+ may be much smaller.

We shall write our solution for the current growth in two forms. One will be valid at all times. The other will only be valid when t/T is a fairly small fraction, say $\frac{1}{4}$ (and it should be noted that in case (a) this t may, nevertheless, be very large compared with T_+). The expression which we shall find for the current growth in this range of small t/T will, of course, be of limited interest; in fact, its chief use will probably be for finding how long a time elapses before a given diffusion process has a significant effect on current growth. In deriving this formula, valid only when t/T is a small fraction, we note that at such times $i_-(t)$ is not very different from what it would be if d were infinite; we may say that by increasing d to infinity we should add a range too far from the cathode to produce much effect there at such times. Thus, the solution valid only in this time-range will consist of the solution for d infinite (with a correction term due to d having its actual value). We proceed now to derive this formula, for t/T a small fraction, and also the exact solution valid at all times.

In the treatment of other secondary processes, Davidson (1953–1957) began by writing the solution as a Laplace contour integral, since from this all possible algebraical forms of the solution can be immediately seen and written down in detail. A similar method will be employed in the present cases. Let the gas be free from electrons and active particles up to the time, $t = 0$, at which the externally generated cathode current I_0 is established. In the boundary conditions (4) and (6) we shall take $h = H = 0$, since the calculation is easily modified to take account of the more general boundary conditions. Then concerning the quantities $n(x, t)$ and $i_-(t)$ we know that

> (i) at $t < 0$, $n(x, t) = i_-(t) = 0$;
> (ii) at $t > 0$, $i_-(t) = I_0 + g_1 D \partial n(x, t)/\partial x$ at $x = 0$;
> (iii) at $t > 0$, $n(0, t) = n(d, t) = 0$; and
> (iv) at $t > 0$, the differential equation (2) holds throughout the gas.

$$(13)$$

These four conditions are sufficient to determine $n(x, t)$ and $i_-(t)$ at all times. Thus if we find functions $n(x, t)$ and $i_-(t)$ which satisfy all four conditions we have the required solution. It may be readily verified that the following expressions satisfy the four conditions:

$$\frac{n(x,t)}{I_0} = \frac{i\alpha_1}{\pi D} \int_C \frac{z \exp[D(z^2 - \mu^2)t]}{(z^2 - \mu^2)\theta} \llbracket [1 - \exp(-2zd)] \exp(\psi x) \tag{14}$$
$$+ \{\exp[(\psi - z)d] - 1\} \exp(-zx) + \{\exp(-2zd) - \exp[(\psi - z)d]\} \exp(zx) \rrbracket dz$$

$$\frac{i_-(t)}{I_0} = \frac{i}{\pi} \int_C \frac{z(z^2 - \psi^2)[(1 - \exp(-2zd)] \exp[D(z^2 - \mu^2)t] dz}{(z^2 - \mu^2)\theta}, \tag{15}$$

where $\psi = \alpha - D(z^2 - \mu^2)/W_-$,

$\theta = \xi + \{2\delta_1 z \exp[(\psi - z)d] - (2\delta_1 z + \xi) \exp(-2zd)\}$,

$\xi = \{z + \psi\}\{(z - \psi)F - \delta_1\}$,

$F = 1$.

The purpose of introducing the symbol F, whose value is unity, will appear later.

The contour C, which is traversed clockwise, is a quarter of an infinite circle, the centre of the circle being at the origin and the centre of the arc being on the positive real axis. (It will be noted that the integrals are written in a form which makes the integrands free from square roots. If we introduced the Laplace form, say by writing $D(z^2 - \mu^2) = p$, this would not be so.)

We note that C may be deformed into a curve on which $[1 - \exp(-2zd)]/\theta$ may be expanded as a series of which the first term is $1/\xi$. The contribution of this term to (15) is

$$\frac{i}{\pi} \int_C \frac{z(z - \psi) \exp[D(z^2 - \mu^2)t] dz}{\{z^2 - \mu^2\}\{z - (\psi + \delta_1)\}}. \tag{16}$$

For use when t/T has not become a large fraction we have to evaluate this expression, with a correction term, say Δ, obtained from the next term in the expansion. Now it is obvious that in case (*a*) and usually in case (*b*) the expression $i_-(t - x/W_-)$ in the differential equation (2) may with negligible error be replaced by $i_-(t)$; that is we may take W_- as infinite; and thus ψ may be replaced by α. To evaluate (16) we note that at $t > 0$ its value is not altered by extending the contour in both directions to form a semicircle lying to the right of the imaginary axis; it may thus be written as the integral taken down that axis, together with two terms from the enclosed poles at $z = \mu$ and $z = \alpha + \delta_1$. Thus, writing $\alpha_2 = \alpha + \delta_1$, we have

$$\frac{i_-}{I_0} = \frac{2\delta_1\alpha_2 \exp\left[D(\alpha_2^2 - \mu^2)t\right]}{\alpha^2 - \mu^2} + \frac{\alpha - \mu}{\alpha_2 - \mu} - \frac{2\delta_1}{\pi} \exp\left(-D\mu^2 t\right) \int_0^\infty \frac{x^2 \exp\left(-Dx^2 t\right) dx}{(x^2 + \alpha_2^2)(x^2 + \mu^2)},$$

$$(17)$$

the last term being the contribution from the imaginary axis. The last term obviously diminishes continuously in magnitude with time, and even at t infinitesimal its magnitude is less than unity (to be precise, it is then $-\delta_1/(\delta_1 + \alpha + \mu)$, thus making the whole expression unity at t infinitesimal).

For the reason explained above, expression (17) must not be used after t has become more than a fraction of T. This limiting fraction is of order $1/\alpha d$ in cases of interest. Within this range of time expression (17) requires only the correction term Δ mentioned above. It is calculated from the next term in the binomial expansion. When added to (17) it multiplies the first term in that expression by a factor which is approximately

$$1 - [4D\alpha_2^2 t \delta_1 \exp\left(-\delta_1 d\right)]/(\alpha_2 + \alpha).$$

For use at larger times the solution will be written in a form valid at all times. To obtain it we use the fact that the integrand in (15) is an odd function of z. Thus the integral taken round an infinite circle is twice the integral taken round an infinite semicircle lying to the right of the imaginary axis, and is thus, at $t > 0$, twice the integral taken round the contour C. By considering the residues at the poles, it is readily seen that at all positive times i_-/I_0 is the real part of

$$A + \sum \frac{2f\lambda(\lambda^2 - \psi^2)\left[1 - \exp\left(-2\lambda d\right)\right] \exp\left[D(\lambda^2 - \mu^2)t\right]}{(\lambda^2 - \mu^2)(\partial\theta/\partial z)_\lambda}. \qquad (18)$$

As before, we may replace ψ by α. The summation extends over all values λ of z (other than 0 and ψ) which satisfy $\theta(z) = 0$, and which lie on the positive real or positive imaginary axes or in the quadrant bounded by them; and f is a factor which is unity for the poles on the axes and 2 for the poles (if any) inside the quadrant. A, which is the contribution from the pole at $z = \mu$ is a constant which is identical with expression (9), and the character of the λ's, all of which are found to lie on the axes, depends on the sign of this quantity.

If expression (9) is positive there will be an infinite number of λ's situated on the imaginary axis; and in this connection it may be noted that if we write $y = iz$ the function θ is

$$2i \exp(iyd)[\{(y^2 + \psi^2)F + \delta_1 \psi\} \sin yd + \delta_1 y(\cos yd - \exp(\psi d))].$$

There can also be a λ on the real axis, but its value will be less than μ. Thus the expression for i_-/I_0 reduces at very large times to the expression (9). A steady state has then been attained.

If, on the other hand, the expression (9) is negative, one of the λ's is real and greater than μ, and thus contributes to the current a term which increases exponentially with time. In either case the exact solution which we have thus obtained may be reduced to an approximate formula valid at large times by including only the principal terms.

Using similar methods, the expression (14) for $n(x, t)$ may be evaluated if desired; but it is not usually of much interest. The important quantities are $i_-(x, t)$ and $i_+(x, t)$ both of which can of course be immediately calculated from the expression for $i_-(t)$.

As remarked above, the treatment may readily be generalized to include the secondary cathode actions of positive ions and unscattered photons; for if in the boundary conditions (13) we add to the right-hand side of (ii) the usual integral expressions which represent these secondary actions, it will be found that, to satisfy all the conditions, we have merely to replace the quantity F, which was unity in the above calculations, by an expression

$$F = 1 - (\delta/\psi)[\exp(\psi d) - 1] - (\alpha\gamma/\phi)[\exp(\phi d) - 1],$$

where $\quad \phi = \alpha - D(z^2 - \mu^2)/W \quad$ and $\quad (1/W) = (1/W_-) + (1/W_+)$.

Since the integrand in (15) is still an odd function of z, expression (18) remains applicable when subjected to this modification, which will give rise to λ's inside the quadrant.

The treatment may also be extended without difficulty to include the effect of photons, generated in the gas by the destruction of metastable atoms or trapped radiation, and proceeding either directly or by diffusion to the cathode.

To obtain expressions for the $i(t)$ produced, not by a maintained I_0, but by a sudden generation of amount q at the cathode at time zero, we have merely to differentiate our expressions for $i_-(t)$ with respect to t and replace I_0 by q.

It is convenient at this point to consider the papers by Newton and other authors, mentioned in the Introduction. These papers do not consider trapped radiation, but they develop methods for calculating approximately the effect of metastable atoms on current growth. In comparison with the method of complex integration, the mathematical methods employed in these papers are not well suited to the problem. In particular, they do not lead to an explicit expression for the amplitude of any term in the series expansion for the current. Instead, they lead to an infinite set of simultaneous equations, each containing an infinite number of terms. Estimating the amplitudes from such a set of equations is an inexact and unduly laborious process. Moreover, these treatments are based on the assumption that both W_- and W_+ may be regarded as infinite. Thus they fail entirely if the γ and δ effects are in themselves sufficient, or even almost sufficient, to produce ultimate breakdown.

5. Similarity

Finally, there is a point of some interest concerning current amplification in similar systems, that is, in systems obtained by varying the pressure p and the field E so as to keep pd and E/p constant. If we assume that trapped radiation has only slight internal destruction, and that its D, besides being constant in time, has geometrical similarity in the systems and is given by (3) with $l/c\tau$ a small fraction, and l inversely proportional to the pressure, then the following results are easily derived.

In a steady state the presence of secondary action by metastable atoms and trapped radiation does not alter the well-known similarity law concerning current amplification which holds in their absence. The same is true in temporal growth if the secondary action by trapped radiation is absent; but when it is present the law is modified; in particular, if no other secondary action of any kind were present, the time taken to attain a given value of i_-/I_0 would be almost independent of d.

6. Conclusions

By applying the formulae obtained in section 4 to particular gases the rate of current growth, due to the simultaneous action of all possible secondary processes including metastable atoms and trapped radiation, may be calculated numerically when the coefficients which appear in the formulae are known. The calculated rate of growth may then be compared with that measured experimentally, in order to determine whether, in the particular gas the action of metastable atoms and trapped radiation is of importance in producing the current growth. Experimental work on these lines is in progress at Swansea.

The author wishes to thank Professor Llewellyn Jones for suggesting the problem and for his continued interest.

References

Davidson, P. M. Appendix to a paper by Dutton, J., Haydon, S. C. and Llewellyn Jones, F., *Brit. J. Appl. Phys.* **4**, 170, (1953).

Davidson, P. M., *Phys. Rev.* **99**, 1072, (1955).

Davidson, P. M., *Phys. Rev.* **103**, 1897, (1956).

Davidson, P. M., *Phys. Rev.* **106**, 1, (1957).

Engstrom, R. W. and Huxford, W. S., *Phys. Rev.* **58**, 67, (1940).

Fowler, R. G., *Handb. Phys.* **22**, 209, (1956).

Llewellyn Jones, F., *Ionization and Breakdown in Gases.* Methuen, (1957).

Molnar, J. P., *Phys. Rev.* **83**, 933, 940, (1951).

Newton, R. R., *Phys. Rev.* **73**, 570, (1948).

Pidduck, F. B., *A Treatise on Electricity*, p. 472. Cambridge University Press, (1925).

Introduction to Papers 11 and 12
(Harrison and Geballe, 'Simultaneous measurement of ionization and attachment coefficients';
Geballe and Reeves, 'A condition on uniform field breakdown in electron attaching gases')

Up until the time of the paper published by R. Geballe and M. A. Harrison in *The Physical Review* 85, 372, (1952), studies of pre-breakdown discharges in electro-negative gases such as oxygen had paid little attention to the influence of the negative ions produced. The analysis of growth-of-current measurements under static, uniform, electric fields was based on the equation $i = i_0 \exp(\alpha d)$ or on

$$i = \frac{i_0 \exp(\alpha d)}{1 - \gamma\,[\exp(\alpha d) - 1]}$$

if secondary ionisation was important. This was in spite of the fact that F. M. Penning had much earlier (in *Ned. T. Natuurkde* 5, 33, (1938)) discussed the effect of attachment, as well as of electronic and vibrational excitation, and of space charge, on the growth of a discharge between parallel plates. Penning gave an equation equivalent to

$$i = i_0 \left\{ \left(\frac{\alpha}{\alpha - \eta} \right) \left[\exp(\alpha - \eta)d - \left(\frac{\eta}{\alpha - \eta} \right) \right] \right\}$$

where η is the attachment coefficient defined in the same manner as the ionisation coefficient α, and estimated values of the attachment coefficient for oxygen.

The two papers reprinted here give in more detail than in the original paper by Harrison and Geballe the analysis and experimental method used in the authors' growth-of-current measurements in attaching gases. The second paper extends the analysis to cases where secondary ionisation is also important. An interesting point made by Geballe and Reeves is that for attaching gases there exists a limiting value of E/p, at which α/p is very nearly equal to η/p, below which breakdown is not obtained at any value of pd.

The occurrence of electron attachment during the spatial development of a pre-breakdown current has made the analysis, using the equation given earlier or equivalent formulae, of growth-of-current experiments considerably more difficult.

Further complication arises with the additional possibility of the reverse process of electron detachment, as has been recognised by Prasad and Craggs (see reference below) and others. More recently it has been recognised that electron detachment may not be the only ion-molecule collision process affecting the growth of pre-breakdown currents. Ion conversion processes in which the original negative ion is converted to an ion of a different species may also be important. The effects of such reactions have been discussed by Moruzzi and Wagner in the papers listed below. As the number of possible processes increases it obviously becomes increasingly difficult to determine the various reaction coefficients with a high degree of accuracy and alternative methods are being sought. For example, much useful information can be obtained by studying mass-spectrometrically the relative abundance of the various ions produced in pre-breakdown discharges. The work described by Kinsman and Rees is an example of the application of this technique.

Further Reading

Penning, F. M., *Ned. T. Natuurkde*, **5**, 33, (1938).

Prasad, A. N., *Proc. Phys. Soc.*, **74**, 33, (1959).

Prasad, A. N. and Craggs, J. D., *Electronics Letters*, **1**, *118*, (1965).

Moruzzi, J. L., *J. Phys. D*, **1**, 1587, (1968).

Wagner, K. H., *Z. Physik*, **241**, 258, (1971).

Kinsman, P. R. and Rees, J. A., *Int. J. Mass. Spectrom. Ion Phys.*, **5**, 71, (1970).

11

Simultaneous measurement of ionization and attachment coefficients

MELVIN A. HARRISON and
RONALD GEBALLE*

Introduction

We have reported previously[1] on simultaneous measurements of the ionization and attachment coefficients in oxygen. Since that time two additional papers[2,3] describing measurements of attachment in oxygen have appeared, and we have extended our procedure to air, CCl_2F_2 (freon-12), and CF_3SF_5. It is the purpose of this paper to relate in more detail the method we have employed, to present our results in these gases, and to discuss these results in comparison with those of other workers.

Following Townsend[4] it has been generally assumed that the current in a pre-breakdown discharge between plane parallel electrodes is described for sufficiently small electrode separation by the familiar equation,

$$i = i_0 \exp{(\alpha d)}, \tag{1}$$

where i is the total current, i_0 the initial electron current at the cathode, d the electrode separation. The ionization coefficient α is the mean number of ionizations per centimetre in the field direction produced by electron collision. Although in complex gases more than one type of ionizing process may occur, no attempt has been made as yet to separate out the corresponding contributions to α. At large separations, as is well known,[5] significant deviations from equation (1) are observed. These are explained in terms of secondary ionizing processes which take place at the cathode or within the body of the gas. Such processes cause a more rapid increase of current with electrode separation than is indicated by equation (1) and are essential to the occurrence of breakdown.

In complex gases and in those containing electronegative atoms other reactions result from electron collision. Ions which are produced in such processes affect the density of current carriers in a manner quite different from the ionization processes described above. Many types of ion-producing reactions have been

* Reproduced from *The Physical Review*, **91**, 1, (1953) by permission of the American Physical Society, New York.

observed, principally through the use of mass spectrographs. At present we confine ourselves to three, designated as follows:

(A) Direct attachment, $AB + e = AB^-$.

(B) Dissociative attachment, $AB + e = (AB^-)^* = A + B^-$.

(C) Dissociation into ions, $AB + e = A^+ + B^- + e$.

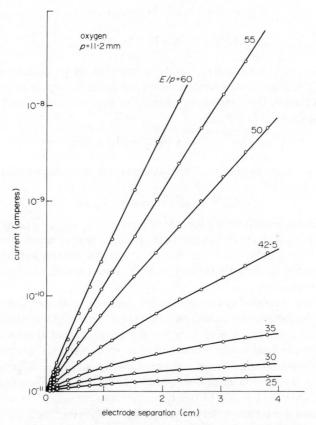

Figure 1. Typical variation of ln i with d in oxygen at 11·2 mm pressure. Each curve gives the results of a run at constant E/p. Data are normalized to the same i_0. Solid curves are drawn from equation (5).

We assume that at a given value of mean electron energy, characterized by the experimental parameter E/p, where E is the field strength in volts/centimetre and p the pressure in millimetres of mercury, one of the reactions predominates and the others can be neglected. Furthermore we permit only those electrode separations which are small enough that secondary ionizing processes of the kind leading to breakdown have no appreciable effect. Focusing our attention on reactions of the types A and B, we describe the attachment process by a parameter η which we define in analogy to α as the mean number of attachments per centimetre of drift in the field direction. In order to find the current which flows when this process and ionization by collision occur simultaneously, we make use of equations of continuity as follows:

$$\partial n_e/\partial t + v_e \partial n_e/\partial x = \alpha v_e n_e - \eta v_e n_e, \tag{2}$$

$$\partial n_+/\partial t + v_+ \partial n_+/\partial x = \alpha v_e n_e, \tag{3}$$

$$\partial n_-/\partial t + v_- \partial n_-/\partial x = \eta v_e n_e. \tag{4}$$

In these equations the subscripts e, $+$, and $-$ refer, respectively, to electrons, positive ions, and negative ions, while n means numerical density, and v stands for drift speed in the gas. The steady-state solution for the total current, with suitable boundary conditions, is found to be

$$i = i_0 [\alpha/(\alpha - \eta)] \exp [(\alpha - \eta)d] - i_0 \eta/(\alpha - \eta), \tag{5}$$

as previously given, When reaction C predominates, the appropriate equations have the solution

$$i = i_0 [(\alpha + \lambda)/\alpha] \exp (\alpha d) - i_0 \lambda/\alpha, \tag{6}$$

where λ is a coefficient describing the rate of reaction C in the same manner that η describes the rates of A and B. Solutions can be readily written down for situations in which reactions B and C occur together and in which secondary ionization processes described by the usual coefficient γ are present as well, but these are of no present concern.

Ordinarily α is measured by determining the slope of a semilogarithmic graph of i against d, on the assumption that equation (1) is correct.[4, 6] Experimental evidence indicates that this is valid procedure for a large number of gases. It has also been applied to gases in which reactions of types A, B and C are known to occur.[7, 8] Now semilogarithmic graphs of equations (5) and (6) are distinctly non-linear. However if α is much larger than η and/or λ, the deviation from linearity will be difficult to observe. A similar difficulty occurs when the range of d is limited to large values, even if α is only slightly larger than η. In this event the additive terms in the equations are negligible, and the coefficient actually deduced will be $\alpha - \eta$. If d, because perhaps of instrumental limitations, is confined to small values (a few millimetres), the deviation again will be hardly apparent. The slope of a straight line drawn through the experimental points in this case will have no simple interpretation. It is apparent from this discussion that only if an extensive range of d is covered by experiment can the extent to which attachment and dissociation affect the discharge current be fully determined, and true values of the coefficients deduced.

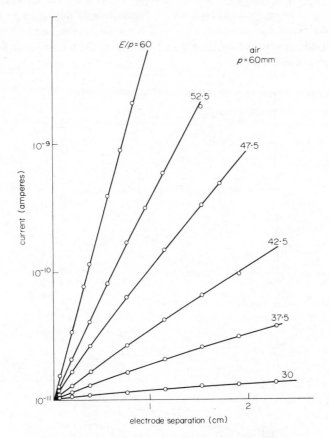

Figure 2. Typical variation of ln *i* with *d* in air at 60 mm pressure
with normalized i_0. Solid curves are drawn from equation (5).
Multiplication is much greater than in oxygen because of the
smaller attachment rate.

Apparatus and procedure
The electrodes are polished copper disks 9 cm in diameter, with contoured edges.
The anode is perforated at its centre with about 400 holes within a circle of 2 cm
diameter, and is rigidly mounted to the Pyrex vessel. A carefully made nut and
screw with 40 threads per inch permit the cathode to be separated any distance
between 0 and 4 cm through the agency of an iron armature which can be rotated
by an external magnet. Ultraviolet light is admitted through a quartz window
directly above the anode and falls normally on the cathode. The light intensity is
monitored by a photocell; occasional adjustment of the load resistor is necessary
to maintain sufficient constancy. The Pyrex vessel is made from a 5-litre flask and
provides a minimum clearance of 4·5 cm. A coating of Dag on the interior wall of

the vessel insures an equipotential boundary. Connection of this coating to a voltage divider permits adjustment of the wall potential. It was found possible to keep current diffusing to the wall less than 1 percent of the total even at the greatest electrode separation.

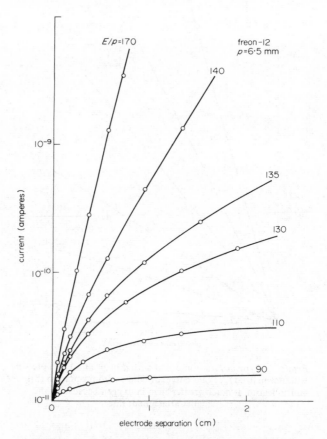

Figure 3. Typical behaviour of ln *i* in freon-12 at 6·5 mm pressure for various values of E/p. Solid curves are drawn from equation (5). Curvature is much more pronounced than in oxygen or air due to the very large attachment coefficient.

The vessel and electrodes have been thoroughly cleaned and baked at temperatures exceeding 400°C in vacuum and an atmosphere of hydrogen for several hours whenever it appeared desirable. Occasionally when either the electrodes became visibly coated or the initial photocurrent was found to be smaller than normal, the electrodes were removed and mechanically cleaned, and the baking process was repeated. The apparatus has not been used unless a pressure of less than 10^{-5} mm of mercury could be maintained for several hours in the closed-off vessel.

In order to be sure that the effects to be reported were not instrumental, measurements were also made in two gases, hydrogen and nitrogen, which are free of any reactions involving ions except ionization by collision.[9-14] Semilogarithmic graphs of current were linear within the accuracy of the measurements, except, that under certain conditions a slight concavity to the electrode separation axis was

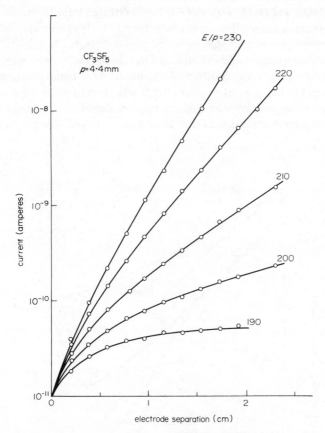

Figure 4. Typical variation of ln *i* with *d* in CF_3SF_5 at 4·4 mm pressure. Solid curves are drawn from equation (5). The saturation at low *E/p* is particularly apparent here.

noted at separations less than three turns of the screw. This effect was found as well when the photocurrent was measured in vacuum. Similar observations have been reported[15] and explained in terms of the reflection of light back up through the holes in the anode. The effect seems definitely correlated with the degree of parallelism of the electrodes; when they are visibly nonparallel, no concavity is observed for these gases. It has been found possible to correct for this effect, utilizing the results of vacuum measurements. The correction is about 3 percent at one turn, 1 percent at three turns.

Attempts were made to obtain considerable purity of all the gases used.

Hydrogen was admitted through a palladium thimble, and in addition other samples of tank hydrogen were passed slowly over hot copper and through a liquid nitrogen trap. Behaviour of hydrogen treated in these various ways was identical. Nitrogen for use in measurements was generated by decomposition of NaN_3, and was passed through a liquid nitrogen trap. Oxygen was produced in three ways: by decomposition of HgO, $KMnO_4$, and MnO_2 with suitable liquid nitrogen traps to remove impurities. Again no difference among them could be detected. Air from outside the laboratory was filtered through a plug of glass wool and passed through a liquid nitrogen trap. Freon came from a commercial cylinder and was distilled in vacuum several times before admission from a trap cooled to the temperature of liquid nitrogen. CF_3SF_5 obtained from the Department of Chemistry[16] was purified in the same manner as freon. Gas pressures were measured with a mercury manometer. During all manipulations except admission of the last two gases a liquid nitrogen trap served

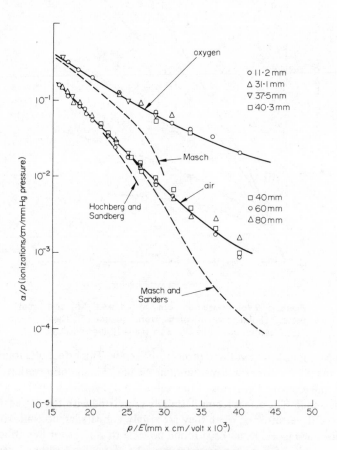

Figure 5. Variation of ionization coefficient with p/E for oxygen and air. Solid curves are drawn by inspection through data of the present work.

as protection against mercury contamination. Subsequent checks indicated that no detectable contamination was admitted in these latter operations. Samples of all gases were changed frequently. At no time was there evidence that a gas had altered its properties while measurements were being made.

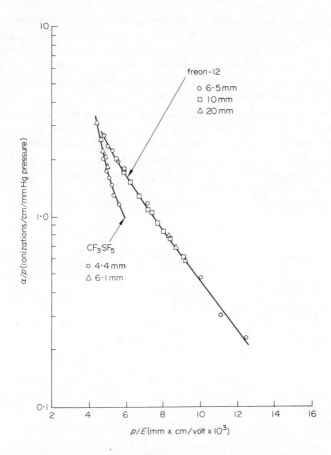

Figure 6. Variation of ionization coefficient with p/E for freon-12 and $CF_3 SF_5$. Solid curves are drawn by inspection through data of the present work.

High voltage was supplied by a regulated negative supply, and was read with a calibrated resistor and a potentiometer. Ionization currents were passed through one of a set of calibrated resistors which were frequently checked against each other. The potential drop across the resistor in use was read with a Dolezalek electrometer. Electrode separations were determined by counting revolutions of the screw. Current measurements were made for varying separations at constant E/p. The gas was never permitted to break down. At sufficiently large E/p, when the electrode separation was increased to more than about 2 cm, a conditioning of the cathode surface appeared to take place which affected the initial photocurrent. To eliminate this effect, measurements in this range of E/p were carried out at either reduced pressure or reduced light intensity.

Data

Samples of data obtained in the gases are shown in *Figures 1* to *4*. The solid curves in these figures are computed from equation (5). By a careful curve-fitting procedure values of α and η can be found from each curve. The procedure fails when at low E/p it is found that $\eta \gg \alpha$, and the current hardly rises above i_0. At high E/p the magnitudes of η and α are reversed, the curvature is slight, and while α can be determined with considerable accuracy, η cannot.

It can be seen that a straight line drawn through the points for separations of greater than one centimetre will pass reasonably close to all of them, and the deviations could be taken for experimental scatter. A straight line could similarly be drawn through the points, say, between 0·2 and 0·8 cm. Neither of these gives the correct α in the region of E/p where detectable curvature exists.

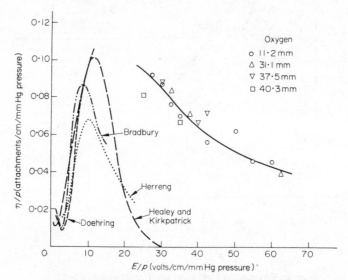

Figure 7. Dependence of attachment coefficient on E/p for oxygen. Solid curve drawn by inspection through present data. Comparison data are given directly in reference 24 but have been computed from the results of references 2, 3, and 23 as discussed in the text.

Data of this nature were taken in oxygen at pressures (measured at 20°C) from 11·2 to 40·3 mm; in air, from 40 to 80 mm; in freon-12, from 6·5 to 20 mm; and in CF_3SF_5, at 4·4 and 6·1 mm. During a run at constant E/p individual points, particularly those at small separations, were repeated to check their reproducibility. In each gas the range of E/p extended throughout the region of applicability of the method for these pressures.

Results and discussion

1. *Measurement of* α/p *and* n/p

Several factors have led us to base our analysis of the curves on equation (5) rather than equation (6). Important among them is the apparent saturation exhibited by curves of *Figures 1* to *4*. Such behaviour is consistent only with the first of these equations. Furthermore in oxygen and in many complex gases it has been found that reactions of type B appear at electron energies of but a few volts,[17–20] whereas those of type C require more than ionization energy. Since dissociative attachment results from resonance capture of an electron, the cross section for this process can attain large values; e.g., a maximum of 10^{-16} cm^2 has been reported recently for SF_6.[20] Under conditions of the present experiments where ionization is just assuming a dominant role it seems quite unlikely that C should occur to an appreciable extent.

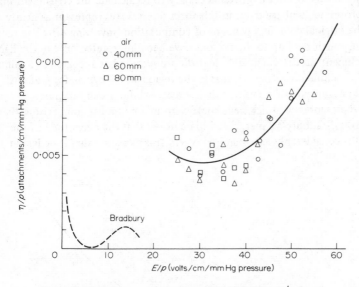

Figure 8. Dependence of attachment coefficient on E/p for air. Solid curve again drawn by inspection. Scatter is greater than for oxygen because of the smaller curvature in *Figure 2*. Comparison curve computed from the results of reference 23. The considerable difference in magnitude has as yet no explanation.

On this basis *Figure 5* shows values of α/p found for oxygen and air plotted against p/E. This scale of abscissas is chosen because it produces for many gases a nearly linear graph. The solid curves are drawn by inspection. Included for comparison are results of other investigations.[8, 21, 22] It should be noted that the measurements of Masch[8] and of Hochberg and Sandberg[22] were made with separations between 0·2 and 0·8 cm, while Sanders[21] did not extend below one centimetre. The first of these authors remarked on the probable effect of attachment on his oxygen data but did not analyze it. In air, where the effect of attachment is considerably diluted, curvature of the semilogarithmic graphs is slight for all E/p. Agreement between results of the quite different measurements of Masch and Sanders in this gas is readily accounted for by this circumstance.

There are no published data with which to compare the curves of *Figure 6* directly. Hochberg and Sandberg[7] have given values of α/p in several complex gases containing electronegative atoms, measured with separations of less than one centimetre. In agreement with their work and consistent with the high dielectric strength of such gases we find the net electron multiplication to be appreciable only at values of E/p considerably larger than for simple gases. The magnitude of α/p however is larger by about one order than reported by these authors.

Figure 7 shows the behaviour of the attachment coefficient deduced for oxygen, reported in part previously.[1] Included for comparison are results of earlier investigations[23, 24] as well as two recent studies.[2, 3] As not all of these authors presented their results in this form, it has been necessary to recalculate the coefficient making use of appropriate auxiliary data. In this form these curves represent as nearly as possible the original data, independent of computations involving drift and random speeds. Since their data up to the present have been interpreted only in the light of an attachment process, incipient electron production has appeared as a falling off of the attachment coefficient. It is in the region where α/p and η/p are of comparable magnitude that three of their curves exhibit a sharp decline.

Figure 8 presents the attachment coefficient in air and, for comparison, that deduced from Bradbury's paper.[23] It will be noted that the magnitude of the coefficient is considerably less than in oxygen. It is expected that it be less in

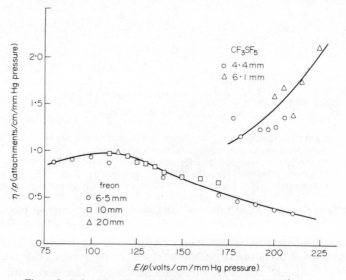

Figure 9. Behaviour of the attachment coefficient with E/p in freon-12 and $CF_3 SF_5$. Curves are drawn by inspection. Scatter in the $CF_3 SF_5$ is principally because these were the first data to be taken.

approximately the ratio of the mole fractions of oxygen in the two gases, in other words, that the peak in air should be about 0.02 attachments per cm per mm. It seems reasonable that such a value would be attained at higher E/p than could be reached by the present apparatus. The positive slope over most of this curve as contrasted with the negative slope observed in oxygen is attributed to the differing energy scales of the two gases. Our data seem to indicate, within the precision of the method, a slight upturn below $E/p = 30$. We are hesitant about the validity of this feature, particularly since the curve should in some manner join that of Bradbury. Whereas in oxygen, the coefficients measured in other researches rise to a magnitude comparable with that found in the present work, this is by no means true in air. For this we have as yet no explanation.

In *Figure 9* are given the attachment coefficients for freon-12 and CF_3SF_5 as functions of E/p. Their magnitudes are comparable with that reported for iodine,[25] and considerably larger than that found in chlorine[26] or expected for fluorine. The data in CF_3SF_5, which were taken early in the work show considerable scatter, and the curve is sketched only to indicate a possible behaviour for this gas. The large magnitude of η/p in these substances might reflect the possibility for multiple reactions of type B. Here again it is apparent that the large probability for attachment could cause a serious underestimate of the rate of electron multiplication.

It is of some interest to point out that in the range of pressures employed no dependence on this parameter has been detected for any of the gases studied.

2. *Dependence of mean cross section on mean energy*

In oxygen and in air it is possible to make further calculations. Auxiliary data exist from which the variation of mean cross section for negative ion production can be correlated with mean electron energy. The cross section is related to η/p by the equation $\sigma = (\eta/p)\,(v_e/c)\,(1/n_0)$, where n_0 is the number of molecules per cubic centimetre at a pressure of one millimetre of mercury, v_e is the drift speed of the electrons, and c their mean speed of agitation. The mean energy is obtained from diffusion measurements in a manner due originally to Townsend.[27] Several sets[24, 27-30] of data of this kind in oxygen and air have been reported, with considerable discrepancy among them. The discrepancies are relatively unimportant in the calculation of cross sections, as only the ratio of the above speeds enters; hence the magnitudes of σ are probably correct in the two gases. As the energy scales (variation of mean energy with E/p) reported by these authors differ widely, it has been necessary to adopt one for each gas on an almost arbitrary basis. In the case of oxygen the data of Brose[29] have been used. Townsend and Tizard[27] present the most complete data for air, but the recent work of Huxley and Zaazou[30] is probably more reliable. Since the latter authors give no data for $E/p > 25$, an extrapolation has been made.

Results of these computations are given in *Figures 10* and *11*, for the cross sections compiled from our results and, for comparison, the work of the other investigators. All curves have been plotted using values of η/p from *Figures 7* and *8* and with auxiliary data as indicated in the previous paragraph. In oxygen the

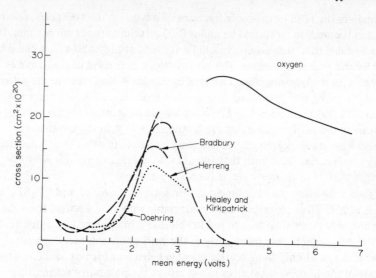

Figure 10. Mean cross section for attachment in oxygen plotted against mean electron energy. All curves are calculated from those of *Figure 7* using the auxiliary data of reference 29.

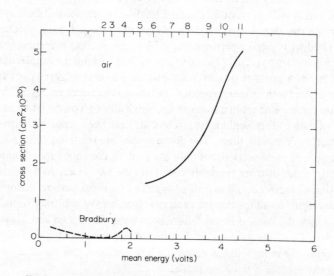

Figure 11. Mean cross section for attachment in air plotted against mean electron energy. The scale at the bottom refers to the work of Townsend and Tizard. That at the top is due to Huxley and Zaazou, although energies above 6 volts are obtained by extrapolation.

maximum agrees in magnitude with the estimate of Massey, 80×10^{-20} cm^2 for the type B reaction, if one recalls that the present value is an average over the energy distribution. Since this author's estimate of the (unaveraged) cross section for the type C reaction is smaller by a factor of almost 4, additional weight is added to the reasons for adopting B as the principal process. The maxima in oxygen and air again should be nearly in the ratio 5:1, deviation from this value being caused by differing energy distributions. Present data are not inconsistent with this conclusion. The increase of cross section as the mean energy approaches thermal values has been attributed to a reaction of type A.[31-33] The peak found at somewhat higher energy by several workers (see *Figure 10*) has been correlated with a level of the oxygen molecule at 1·62 V.[23, 34] Inelastic collisions at this energy, it has been suggested, return electrons to the very low energy region where they are captured by reaction A. Hagstrum[35] and others,[17] however, have shown that reaction B has an appearance potential of 6 V and a peak probability at 8 V. The previously mentioned estimate for the height of this peak (80×10^{-20} cm^2) provides sufficiently well for the presently observed maximum to indicate that dissociative attachment is responsible in large part for the observed behaviour of the oxygen. Since the excitation function for the 1·62-V level is presently unknown, the contribution of this 'reflection' process cannot be estimated.

The authors wish to acknowledge helpful comments by interested friends, in particular those of Professor L. B. Loeb of the University of California.

References and Notes

1. Geballe, R. and Harrison, M. A. *Phys. Rev.* **85**, 372, (1952).
2. Doehring, A., *Z. Naturforsch.* **7a**, 253, (1952).
3. Herreng, P., *Cahiers phys.* **38**, 7, (1952).
4. Townsend, J. S., *Nature* **62**, 340, (1900).
5. Loeb, L. B., *Fundamental Processes of Electrical Discharge in Gases*, John Wiley and Sons, Inc., New York, (1939), Chap. IX.
6. Reference 5, Chap. VIII.
7. Hochberg, B. M. and Sandberg, E. J., *Compt. rend. acad. sci. U.R.S.S.* **53**, 511, (1946).
8. Masch, K., *Arch. Elektrotech.* **26**, 587, (1932).
9. Mass spectrographic investigations have produced no evidence for the existence of N^- or N_2^- (see references 10 and 11). H^- has been reported in very small amounts (see references 12 to 14) but no estimates of the cross section for its formation are given. The present apparatus would have detected a process with a mean cross section of 10^{-21} cm^2.
10. Tate, Smith, and Vaughan, *Phys. Rev.* **48**, 525, (1935).
11. Tate, J. T. and Lozier, W. W., *Phys. Rev.* **39**, 254, (1932).
12. Lozier, W. W., *Phys. Rev.* **36**, 1417, (1930).
13. Smyth, H. D. and Mueller, D. W., *Phys. Rev.* **43**, 116, (1936).
14. Tüxen, O., *Z. Physik* **103**, 463, (1936).

15. Kruithoff, A. A. and Penning, F. M., *Physica* 3, 515, (1936).
16. Silvey, Gene A. and Cady, George H., *J. Amer. Chem. Soc.* 72, 3624, (1950).
    We wish to thank Dr. Silvey and Professor Cady for supplying this compound.
17. Lozier, W. W., *Phys. Rev.* 46, 268, (1934).
18. Baker, R. F. and Tate, J. T., *Phys. Rev.* 52, 683, (1938).
19. Vought, R. H., *Phys. Rev.* 71, 93, (1947).
20. Ahearn, A. J. and Hannay, N. B., *J. Chem. Phys.* 21, 119, (1953).
21. Sanders, F. H., *Phys. Rev.* 41, 667, (1932); 44, 1020, (1933).
22. Hochberg, B. M. and Sandberg, E. J., *J. Tech. Phys.* (*U.R.S.S.*) 12, 65, (1942).
23. Bradbury, N., *Phys. Rev.* 44, 883, (1933).
24. Healey, R. H. and Reed, J. W., *The Behavior of Slow Electrons in Gases*, Amalgamated Wireless, Sydney, (1941), p. 94.
25. Healey, R. H., *Phil. Mag.* 26, 940, (1938).
26. Bailey, V. W. and Healey, R. H., *Phil. Mag.* 19, 725, (1935).
27. Townsend, J. S. and Tizard, H. T., *Proc. Roy. Soc.* A87, 357, (1912); A88, 336, (1913).
28. Townsend, J. S. and Bailey, V. A., *Phil. Mag.* 42, 873, (1921).
29. Brose, H. L., *Phil. Mag.* 50, 536, (1925).
30. Huxley, L. G. H. and Zaazou, A. A., *Proc. Roy. Soc.* A196, 402, (1949).
31. Bloch, F. and Bradbury, N., *Phys. Rev.* 48, 689, (1935).
32. Biondi, M., *Phys. Rev.* 84, 1072, (1951).
33. Massey, H. S. W., *Negative Ions*, Cambridge University Press, Cambridge, (1950), second edition, p. 72.
34. Bates, D. R. and Massey, H. S. W., *Trans. Roy. Soc.* (London) A239, 269, (1943).
35. Hagstrum, H. D., *Rev. Modern Phys.* 23, 185, (1951).

12

A condition on uniform field breakdown in electron-attaching gases

RONALD GEBALLE AND MARVIN L. REEVES*

The general form of the curve that relates breakdown potential to the product of electrode separation and pressure pd in gases between plane-parallel electrodes is well-known.[1] A curve of this kind can be converted into one relating E/p for breakdown (where E is the electric field strength) and pd, assuming uniformity of field. As pd increases, the latter curve falls rapidly at first, but eventually more slowly and seems, where it has been carried sufficiently far, to approach an asymptotic value below which breakdown does not occur.[2,3] It is the main purpose of the present note to suggest an interpretation of this limiting E/p for breakdown in gases that form negative ions by electron attachment.

Since the existence of an asymptote seems common to all gases its explanation would appear to depend on the mechanism of breakdown. In particular it might be related to the failure of some secondary ionizing process at low E/p. However, the breakdown mechanism is incompletely understood at present, and it is difficult to make even a qualitative statement. In spite of this limitation the characteristics of attaching gases permit certain conclusions to be drawn.

An equation for steady-state pre-breakdown current in attaching gases has been presented previously.[4] That equation is appropriate when secondary ionizing processes such as photoelectric effect or positive ion bombardment are unimportant. Discussions of breakdown, however, must include such processes. Assuming a secondary process at the cathode, one finds

$$i/i_0 = \frac{[\alpha/(\alpha - \eta)] \exp [(\alpha - \eta)d] - \eta/(\alpha - \eta)}{1 - \{\gamma\alpha/(\alpha - \eta)\} \{\exp [(\alpha - \eta)d] - 1\}}, \tag{1}$$

where the symbols are to be interpreted as follows: i = total discharge current, i_0 = electron current produced at the cathode by an external source, α = number of ionizations/cm electron = the ionization coefficient, η = number of attachments/cm

* Reproduced from *The Physical Review*, **92**, 867, (1953) by permission of The American Physical Society, New York.

electron = the attachment coefficient, d = electrode separation, and γ = probability per positive ion or photon of liberating an electron from the cathode.

A threshold for a self-sustaining current is obtained by requiring the denominator of equation (1) to vanish, i.e.,

$$\{\gamma\alpha/(\alpha - \eta)\}\{\exp[(\alpha - \eta)d] - 1\} = 1. \tag{2}$$

It is well-known[5] that for non-attaching gases ($\eta = 0$) equation (2) can be fitted to observed breakdown potentials within experimental error. In addition it has been shown recently[6] that a nearly equivalent procedure can be used to represent formative times of sparks at low overvoltages. If we suggest that equation (2) is equally valid for breakdown in attaching gases we note that when $\alpha/p \geqslant \eta/p$, breakdown is possible for sufficiently large pd regardless of the values of α/p, η/p, and γ. Under this condition the breakdown criterion always has a pd dependence. However for $\alpha/p < \eta/p$ equation (2) approaches, with increasing pd, an asymptotic form which is independent of pd, i.e.:

$$\alpha/p = (\eta/p)/(1 + \gamma). \tag{3}$$

Equation (3) is a condition on E/p alone and fixes a limit for this parameter, $(E/p)_{\lim}$, below which no sparking should be possible regardless of the magnitude of pd. In the following we shall compare, for several gases, values of $(E/p)_{\lim}$ obtained from these considerations with asymptotic limits derived from experimental studies of breakdown potentials.

Values of α/p and η/p are now available in a number of attaching gases.[4] In all cases thus far α/p is the smaller at low E/p but grows more rapidly with increasing E/p and overtakes η/p. Values of E/p for equality vary widely from gas to gas, ranging from 30 to 300 volts/cm/mm. They are believed accurate within 5 percent.

Values of γ will vary with gas and electrode material. Its order of magnitude has been found in air[7] to be less than 10^{-3} near the asymptotic E/p for this gas. At high E/p, Hale,[8] using H_2 with electrodes of Pt and Na, found γ to lie below 0.05 for E/p below 600 except for a peak of no present concern. The magnitude of γ is expected, in general, to be sufficiently small that this quantity can be neglected in equation (3), a simplification further justified by uncertainties in the values of $(E/p)_{\lim}$ available from experiment. Thus, we are led to adopt the relation

$$\alpha/p = \eta/p \tag{4}$$

as a working condition for the limiting E/p.

Comparison with experiment can be made through equation (2). The substitution of values of α/p and η/p yields, as for the non-attaching case, a relation between breakdown potential and pd with one parameter to be determined. In CCl_2F_2 and oxygen, for which the most extensive checks can be made at present, experimental curves are duplicated as closely as those of non-attaching gases. However, this is at best an insensitive test.

A direct test of equation (4) is illustrated in *Table 1*. Here values of E/p_{lim} predicted from this equation by the use of data from reference 4 are compared with those deduced from measurements of breakdown potentials. It is important to

Table 1—Comparison of Predicted and Observed $(E/p)_{lim}$

| | $(E/p)_{lim}$ | | % | Dielectric | |
Gas	Pred.	Obs.	dev.	strength	Ref.
CCl_4	305	294	3	6	a
CF_3SF_5	186	160	15	3	b
CCl_2F_2	126	110	14	2·4	c
SF_6	117	103	13	2·2	d
Air	31·5	36·5	15	1	e
O_2	35·5	36·5	3	0·95	f

a Hochberg, B. M. and Sandberg, E. J., *J. Tech. Phys. (U.S.S.R.)* **12**, 65, (1942).
b Geballe, R. and Linn, F. S., *J. Appl. Phys.* **21**, 592, (1950).
c Trump, Safford, and Cloud, *Trans. Am. Inst. Elec. Engrs.* **60**, 132, (1941).
d Wilson, Simons, and Brice, *J. Appl. Phys.* **21**, 203, (1950).
e Fisher, L. H., *Phys. Rev.* **72**, 423, (1947).
f Fisher, L. H., (private communication).

point out two principal causes of error in the latter. If measurements are not made under conditions of strict uniformity of electric field, breakdown will occur at abnormally low E/p. On the other hand, if measurements are not carried out to sufficiently large pd, the limiting E/p cannot be estimated accurately and will generally be given too high a value. Best conditions of uniformity seem to have been achieved in CCl_2F_2, oxygen and air. Only for the first two of these do measurements permit of reliable extrapolation to the asymptotic limit. It should be pointed out, in addition, that a failure of the kind described in the second paragraph might cause the observed $(E/p)_{lim}$ to lie above the predicted value. In view of these uncertainties, the agreement exhibited in *Table 1* indicates the essential correctness of the proposed interpretation of equation (4). It can be seen from the table that an ordering according to $(E/p)_{lim}$ corresponds closely to relative dielectric strength as generally quoted.

The authors are grateful to Professor L. H. Fisher for supplying breakdown potential data in oxygen prior to publication.

References

1. Loeb, L. B., *Fundamental Processes of Electrical Discharge in Gases*, John Wiley and Sons, Inc., New York, (1939), Chap. X.
2. Druyvesteyn, M. J. and Penning, F. M., *Rev. Modern Phys.* **12**, 88, (1940).
3. Loeb, L. B. and Meek, J. M., *The Mechanism of the Electric Spark*, Stanford University Press, Stanford, (1941), p. 74.
4. Geballe, R. and Harrison, M. A., *Phys. Rev.* **85**, 372, (1952); Harrison, M. A. and Geballe, R., *Phys. Rev.* **91**, 1, (1953); Reeves, M. L. and Geballe, R. (to be

published). Note added in proof: Dr. J. D. Craggs has kindly called our attention to an article by Penning, F. M., *Ned. T. Natuurkde* **5**, 33, (1938) in which this equation is derived.

5. See references 1-3.
6. Kachikas, G. A. and Fisher, L. H., *Phys. Rev.* **88**, 878, (1952).
7. Llewellyn Jones, F. and Parker, A. B., *Proc. Roy. Soc.* **A213**, 185, (1952).
8. Hale, D. H., *Phys. Rev.* **55**, 815, (1939).

Introduction to Paper 13 (Bhalla and Craggs, 'Measurement of ionization and attachment coefficients in sulphur hexafluoride in uniform fields'

It is of considerable interest industrially that many polyatomic molecules such as sulphur hexafluoride and various fluorocarbons, e.g. those of the series $C_n F_{2n+2}$, have high dielectric (i.e. breakdown) strengths. Over the last fifteen years a number of these gases have been examined by Professor Craggs and his colleagues, both in single-collision experiments and under swarm conditions. In the single-collision experiments (i.e. experiments done at low gas pressures $\approx 10^{-4}$ torr) the appearance potentials and ionisation, attachment and detachment cross-sections have been examined, while under swarm conditions the ionisation and attachment coefficients have been determined in Townsend growth-of-current experiments. More recently, associated quantities such as the electron and ion drift velocities and mean energies, and ion-molecule reaction rates, have been studied. The paper by Bhalla and Craggs is typical of the growth-of-current experiments and deals with one of the more important gases, sulphur hexafluoride. For details of work carried out with other gases which readily form negative ions one may consult the references given below.

Further Reading
Warren, J. W., Hopwood, W. and Craggs, J. D., *Proc. Phys. Soc.* **B63**, 180, (1950).
Warren, J. W. and Craggs, J. D., Rep. on Conf. on Mass Spect. (London Institute of Petroleum) p. 36, (1952).
Asundi, R. K. and Craggs, J. D., *Proc. Phys. Soc.* **83**, 611, (1964).
Bozin, S. E. and Goodyear, C. C., *J. Phys.* D, **1**, 327, (1968).
Razzak, S. A. A. and Goodyear, C. C., *J. Phys.* D, **1**, 1215, (1968).

13

Measurement of ionization and attachment coefficients in sulphur hexafluoride in uniform fields

M. S. BHALLA AND J. D. CRAGGS*

1. Introduction

Sulphur hexafluoride (SF_6) is one of many compounds of polyatomic molecules that have a dielectric strength appreciably higher than air or nitrogen, but has the advantage that it may be used up to much higher gas pressures than most of the highly attaching gases because of its low critical temperature. In view of this advantage, SF_6 is of considerable importance as a dielectric; therefore an understanding of negative ion formation and its bearing on the breakdown strength of this gas is desirable. The attachment processes occurring in SF_6 when submitted to bombardment by low-energy electrons of well-defined energy, at low pressures (10^{-5}-10^{-4} mm Hg) have been studied by Ahearn and Hannay (1953), Marriott (1954), Marriott and Craggs (1956) by using the conventional mass spectrometer methods and by Hickam and Fox (1956) by applying their retarding potential difference method. The appearance potentials and relative abundances of the different positive ions in SF_6 have also been studied by Dibeler and Mohler (1948) and Marriott (1954) by using the mass spectrometer methods.

Values of the Townsend ionization coefficient α as a function of E/p (E is the field strength, p the gas pressure) have been measured in SF_6 by Hochberg and Sandberg (1942, 1946). Since nothing was known about the negative ion formation in SF_6 at that time no allowances were made for the attachment processes occurring in the gas. Geballe and Reeves (Loeb, 1955) have measured the values of α and of the dimensionally equivalent attachment coefficient η, defined as the number of attachments per electron per centimetre in the field direction, in SF_6 as a function of E/p by the technique of Harrison and Geballe (1953). Further, Geballe and Reeves (1953) have indicated that for SF_6 there is a limiting value of $E/p = 117$ V cm^{-1} mmHg^{-1} below which no sparking should be possible regardless of the magnitude of pd (d is the electrode separation) and this $(E/p)_{\lim}$ is nearly that for which α and η are equal. McAfee (1955) has also measured η in SF_6 for

* Reproduced from *Proceedings of the Physical Society,* **80**, 151, (1962) by permission of The Institute of Physics and The Physical Society, London.

low values of the parameter E/p using a pulse technique. The pressures used by Geballe and Reeves (1953) and McAfee (1955) in their experiments were less than 20 mmHg and 1 mmHg respectively. It therefore seemed desirable to measure the values of α and η in SF_6 for various values of the parameter E/p and to study the breakdown mechanism in this gas at higher pressures in uniform field conditions.

It has been shown (Harrison and Geballe, 1953; Geballe and Reeves, 1953) that when electron attachment of the type

$$AB + e \rightarrow AB^- \quad \text{or} \quad \rightarrow A + B^- \tag{1}$$

occurs in the growth of the electron avalanche, the equation for the current flowing in a uniform field gap is given, in the absence of detachment, by

$$\frac{I}{I_0} = \left[\frac{\alpha}{\alpha - \eta} \exp\left[(\alpha - \eta)d\right] - \frac{\eta}{\alpha - \eta} \right] \Big/ \left[1 - \frac{\gamma\alpha}{\alpha - \eta} \left(\exp\left[(\alpha - \eta)d\right] - 1 \right) \right] \tag{2}$$

where α is Townsend's primary coefficient, η the attachment coefficient, γ the secondary coefficient, d the electrode separation and I_0 the initial electron current at the cathode.

Provided only those current measurements are considered in which contributions due to secondary processes are negligible, then equation (2) is reduced to

$$\frac{I}{I_0} = \left[\frac{\alpha}{\alpha - \eta} \exp\left[(\alpha - \eta)d\right] - \frac{\eta}{\alpha - \eta} \right]. \tag{3}$$

Equation (3) indicates that the semi-logarithmic plot of I/I_0 against d is not linear and that curve fitting techniques are necessary to evaluate α and η accurately.

It follows from equation (2) that the criterion for breakdown at which the current I becomes self-maintained and independent of I_0 is

$$\frac{\gamma\alpha}{\alpha - \eta} \{ \exp\left[(\alpha - \eta)d_s\right] - 1 \} = 1 \tag{4}$$

where d_s is the breakdown sparking distance (cm). When $\alpha/p < \eta/p$, equation (4) approaches an asymptotic form as pd increases, given by

$$\frac{\alpha}{p} = \frac{\eta/p}{1 + \gamma}. \tag{5}$$

This depends only on E/p and fixes a limiting value $(E/p)_{lim}$ below which no sparking should be possible, however large pd may be. If γ is negligible then the limiting condition is simply $\alpha/p = \eta/p$.

Studies by earlier workers, working at low pressures, show that heavy negative ions, including in particular SF_6^- and SF_5^-, are found in SF_6 by resonance capture of very slow electrons at about 0·05 eV. The capture process has been interpreted by Ahearn and Hannay (1953) as

$$SF_6 + e \rightarrow (SF_6^-)* \begin{cases} SF_6^- & \quad (6) \\ SF_5^- + F & \quad (7) \end{cases}$$

$((SF_6^-))*$ denotes an excited state of SF_6^-).

To the authors' knowledge, no information is available concerning the mean electron energies in SF_6. However, from the general electron energy distribution considerations and the available attachment cross section data in SF_6 it appears that the predominant process of negative ion formation, in the range of parameters used in the present study, is due to the capture of very slow electrons producing SF_6^- and SF_5^- (equations (6) and (7)). This mechanism of negative ion formation has been assumed, and consequently values of α, η, γ and the breakdown sparking distance d_s have been calculated in SF_6 by employing equations (2), (3) and (4). The authors find that the measurements of pre-breakdown currents conform to the above theory over the wide range of pressures studied.

2. Experimental procedure

A detailed description of the apparatus and the modifications made to improve its performance and accuracy have been published earlier (Hopwood *et al.*, 1956, Bhalla and Craggs, 1960) and will not be dealt with here. In the present study the ionization chamber (about 80 litres in volume) was always evacuated to better than 10^{-5} mmHg with an apparent leak of less than $0·4 \mu h^{-1}$ before the gas was admitted. Pre-dried compressed SF_6 of 99·98% purity supplied in a cylinder by Imperial Chemical Industries Ltd. was used throughout this study. The gas was however passed very slowly through a trap maintained at $-40°C$, by using a mixture of solid carbon dioxide and acetone as a further precaution. Activated alumina was used as an absorbent (placed inside the chamber) so that the products of dissociation formed in SF_6 by electrical discharges (Schumb *et al.*, 1949) do not attack or corrode the apparatus. The chamber was opened for inspection after the experiments and no effects due to any chemical action on the electrode surfaces, chamber walls or glass parts could be detected. Further, check measurements of α/p were made in hydrogen before and after using SF_6 in the chamber, and showed no detectable differences.

The gas pressures were measured (corrected to $20°C$) either on the oil manometer using silicone oil of specific gravity 1·09 to within $\pm 0·05$ mmHg or on the micro-manometer (cf. Hopwood *et al.*, 1956) to within $\pm 0·5$ mmHg. A platinum-coated cathode was used throughout this study and its distance from the Dural (aluminium alloy) anode was measured to within $\pm 0·005$ mm. The voltage applied to the ionization chamber, variable in steps of 0·2 volt (0 to 50 kV) was known to within 0·3%, and the ionization currents produced could be determined with an accuracy to better than 1·5%. It is very difficult to assess the errors in α/p and η/p since they depend on a curve-fitting technique, but it is estimated that in the most favourable conditions they are about ± 5%, increasing to perhaps ± 15% for low values of E/p.*

3. Data and results

The experiments were conducted over a pressure range of 5 to 200 mmHg (corrected to 20°C) and a range of E/p from 90 to 160 V cm^{-1} mmHg^{-1}. Typical $(\log I, d)$ plots at 5 and 25 mmHg pressures are shown in *Figures 1* and *2* respectively. *Figures 1* and *2* also indicate the 'saturation' currents measured at a gap of 0·5 cm and voltages of a few tens of volts to a few hundred volts (curves A and B) at 5 and

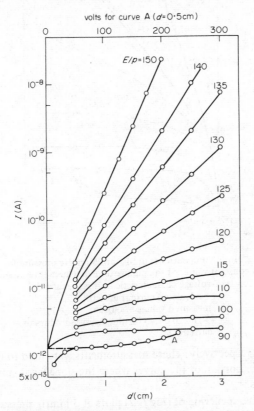

Figure 1. Ionization currents in SF$_6$ at 5 mmHg pressure for various constant values of the parameter E/p. Curve A represents current/voltage characteristic at this pressure.
—, calculated current growth;
o, measured values of current.

* Conditions are most favourable when $(\alpha - \eta)$ is not too large and preferably lies between -1 and $+1$. The size of the term exp $(\alpha - \eta)d$ is important and it should not dominate equation (2). The basis of the estimate is the accuracy of the repetition of independent calculations, and experience with a number of attaching gases has been of value in this respect.

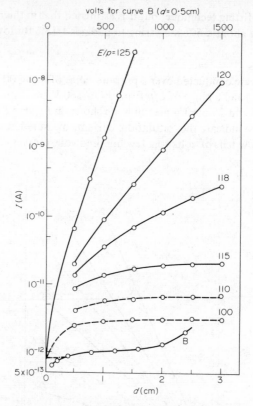

Figure 2. Ionization currents in SF_6 at 25 mmHg pressure for various constant values of the parameter E/p. Curve B represents current/voltage characteristic at this pressure.

—, calculated current growth;
o, measured values of current;
– – –, mean curve for the measured values of current.

25 mmHg pressures respectively. These measurements were used to determine the values of I_0 from the portion of the curves where the slope was a minimum, as shown in the figures.

Figure 3 shows the upcurving of $(\log I, d)$ plots at 5 mmHg pressure, indicating the presence of secondary ionization processes. The initial photoelectric current I_0 from the cathode was reduced by adding a fine wire gauze filter in the path of the ultra-violet radiation from the low pressure mercury-vapour discharge lamp incident on the cathode. This enabled pre-breakdown currents to be measured sufficiently close to the sparking distance without causing space-charge distortion of the electric field due to unnecessarily large values of anode current. It was not possible to take such measurements at 25 and 100 mmHg pressures because the ionization currents became unstable near the sparking distances before any upcurving in the $(\log I, d)$ plots could be observed.

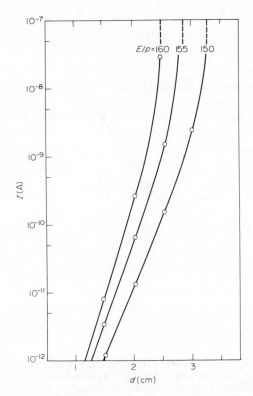

Figure 3. Ionization currents in SF_6 at 5 mmHg pressure for
various constant values of the parameter E/p.

3.1. *Evaluation of α and η*

The values of α and η have been computed from the above-mentioned curves by
a careful process of curve fitting employing equation (3). The values of α/p, η/p and
$(\alpha - \eta)/p$ at different pressures are plotted as a function of E/p in *Figures 4, 5* and *6*
respectively, which also include the data of Geballe and Reeves (Loeb, 1955) for
comparison. *Figure 5* also shows McAfee's data obtained by the pulse technique.

3.2. *Breakdown measurements and γ*

Breakdown potentials (V_s) were measured over a pressure range of 5 to
200 mmHg (corrected to $20°C$) up to $pd \approx 400$ mmHg cm, and are plotted as a
function of pd in *Figure 7*, which also includes Prasad's (1959, 1960) data in dry
air for comparison. Results showed no pressure dependence (within 1%) for any
particular value of pd except for the gaps greater than 2·5 cm at 100 mmHg pressure.
In addition, the values of E/p obtained from the breakdown potentials and gap
distances at different pressures are plotted as a function of pd in *Figure 8*.

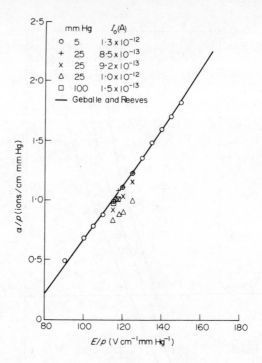

Figure 4. Values of α/p as a function of E/p in SF_6; the effect of using various possible values for I_0 in the calculation of the coefficient from the results at $p = 25$ mmHg is shown.

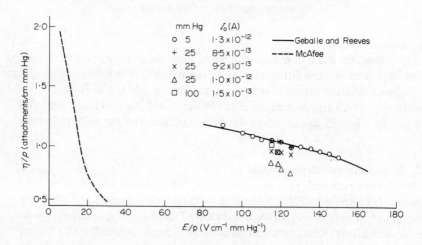

Figure 5. Values of η/p as a function of E/p in SF_6; the effect of using various possible values for I_0 in the calculation of the coefficient from the results at $p = 25$ mmHg is shown.

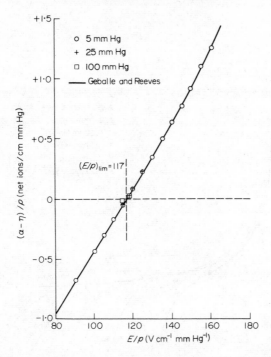

Figure 6. Values of $(\alpha - \eta)/p$ as a function of E/p in SF_6.

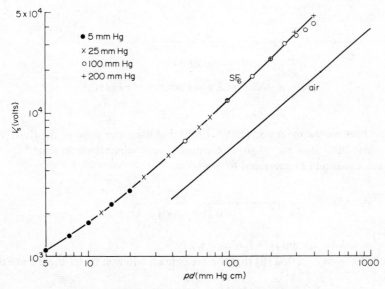

Figure 7. Observed breakdown potentials as a function of pd
in SF_6 and in dry air for comparison.

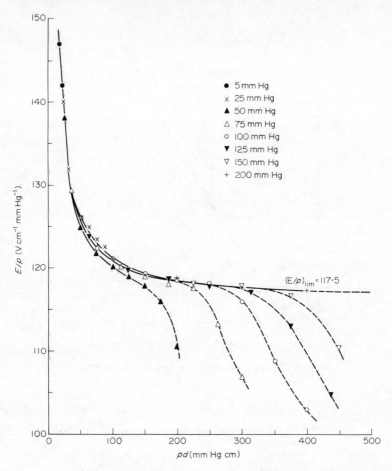

Figure 8. Values of E/p as a function of pd in SF_6.

It has been shown by Prasad (1959, 1960) that the mean slope of the $(\log I, d)$ plots, where they show very slight curvatures, is approximately represented by $\alpha - \eta$ and equation (2) is reduced to

$$\frac{I}{I_0'} = \frac{\exp\,[(\alpha - \eta)d]}{1 - \gamma'\,\{\exp\,[(\alpha - \eta)d] - 1\}} \tag{8}$$

where $\gamma' = \gamma\alpha/(\alpha - \eta)$ and $I_0' = I_0\alpha/(\alpha - \eta)$.

It follows from equation (8) that the breakdown criterion under these conditions is given by

$$\gamma'\,\{\exp\,[(\alpha - \eta)d] - 1\} = 1. \tag{9}$$

Values of the secondary coefficient γ' have been computed from the upcurving of linear (log I, d) plots (cf. *Figure 3*) at 5 mmHg pressure by employing equation (8). These values of γ' have been used to calculate the sparking distances d_s from the breakdown criterion (9). The table gives the values of $(\alpha - \eta)/p$, γ' and calculated d_s with measured d_s for different values of the parameter E/p at 5 mmHg pressure. Calculated values of the sparking distances agree with the measured values to within 1% (see table) showing that the static breakdown in SF_6 at 5 mmHg pressure is brought about by the Townsend build-up mechanism. It was not possible to measure γ' at 25 and 100 mmHg pressures owing to the reasons mentioned earlier (c.f. section 3).

Data relating to Breakdown in SF_6

E/p	$(\alpha - \eta)/p$	Mean γ (meas.)	d_s (cm) (calc.)	d_s (cm) (meas.)
150	0·95	$1·6 \times 10^{-7}$	3·30	3·32
155	1·10	$1·7 \times 10^{-7}$	2·84	2·86
160	1·24	$1·8 \times 10^{-7}$	2·50	2·51

4. Discussion

Stability of the photoemission from the cathode (I_0) is of great importance for reproducibility of the pre-breakdown current measurements. These measurements were repeatedly checked and could be reproduced to within 3% for each gas filling, thereby showing that I_0 was stable, but it is seen from figures 1, 2 and 3 that the value of I_0 decreased approximately exponentially with increase in the gas pressure. This approximately exponential fall in I_0 suggests absorption of the ultra-violet radiation which traverses a distance of approximately 40 cm through the gas before reaching the cathode (Hopwood *et al.*, 1956), but at the same time possibility of the pressure-dependent cathode effects cannot be ruled out. Liu *et al.* (1951) studied the absorption in SF_6, in the vacuum ultra-violet region, and observed slight absorption up to about 2170 Å. However, Nostrand and Duncan (1954) have shown that the absorption is about 1 cm^{-1} at n.t.p. at 1420 Å, but is considerably smaller above this wavelength and is effectively zero near 1870 Å. This indicates that the decrease in I_0 with increase in the gas pressure cannot be attributed to the absorption of the ultra-violet radiation incident on the cathode. If, however, this effect was due to the absorption of ultra-violet radiation from the lamp it would not cause any appreciable change in I_0 when the gap d was varied from 0·5 to 3 cm. Consequently, it appears that normal adsorption of the gas or its dissociated products on the cathode surface increased the average cathode work function and was responsible for the decrease in I_0 with increase in the gas pressure, though the details of the mechanisms involved are not thoroughly understood (see Appendix).

Effects due to the field distortion caused by the holes in the anode (cf. Hopwood *et al.*, 1956) might affect the pre-breakdown current measurements for $d < 0·5$ cm, so that these have not been considered in computing the values of α and η. Moruzzi (private communication) has recently, in this laboratory, carried out experiments which show that field distortion due to the anode holes is negligible for gap lengths greater than 3 mm. Since there was no other method of assessing the value of I_0, 'saturation' current measurements at a gap of 0·5 cm were used to determine the values of I_0 (cf. section 3 and *Figures 1* and *2*). There was no difficulty in assessing the value of I_0 from the curve A at 5 mmHg pressure, but no particular value of I_0 could be chosen from the curve B at 25 mmHg pressure as no portion of the curve was parallel to the voltage axis. Therefore, values of α/p, η/p and $(\alpha - \eta/)p$ have been computed for three different values of I_0 ($8·5 \times 10^{-13}$ A, $9·2 \times 10^{-13}$ A and $1·0 \times 10^{-12}$ A) and are shown in *Figures 4* and *5*. *Figures 4* and *5* also show that the maximum variation in the values of α/p and η/p at any particular value of E/p is not more than 20%. The values of $(\alpha - \eta)/p$ for these different values of I_0 do not differ from each other by amounts greater than 3% (the experimental error) so that they are not shown separately in *Figure 6*.

The values of α/p, η/p and $(\alpha - \eta)/p$ over a wide range of the parameter E/p plotted in *Figures 4*, *5* and *6*, respectively, show good agreement with the earlier measurements of Geballe and Reeves (Loeb, 1955). The values of α/p and η/p do not show any apparent pressure dependence and are therefore functions of E/p. This consideration indicates that the predominant process of negative ion formation in SF_6, in the range of parameters used in this study, is the capture of slow electrons producing SF_6^- and SF_5^- (primary capture process), and that the contribution due to the secondary capture processes, in which the yield of negative ions is proportional to the square of the pressure is negligible. However, it is possible that the latter processes may become important at higher pressures. *Figure 6* also shows McAfee's data (1955), which do not appear to form a continuous curve with the data obtained either from this study or from the experiments of Geballe and Reeves. This discrepancy has been fully discussed by Berg and Dakin (1956) and will not be dealt with here.

Values of the secondary coefficient γ' are very low, i.e. of the order of 10^{-7} (see table). Measured values of the static breakdown potentials in SF_6 are considerably higher than those in dry air for the corresponding values of pd (see *Figure 7*), thereby giving high relative dielectric strength, for example, approximately 2·8 at 380 mmHg cm (see Howard, 1957), as compared with dry air. The values of E/p are plotted as a function of pd in *Figure 8*, which shows that as pd increases the value of E/p falls rapidly at first, then gradually and eventually approaches a limiting value below which breakdown does not occur until pd is increased beyond a value which increases with pressure. The subsequent decrease in breakdown E/p beyond the limiting value set by $\alpha/p = \eta/p(1 + \gamma)$, and which was first described by Geballe and Reeves (1953), has not, to our knowledge, been previously reported. The 'limiting value' of $E/p = 117·5$ V cm^{-1} mmHg^{-1} shows good agreement with the predicted value of 117 V cm^{-1} mmHg^{-1} as shown in *Figure 6*. Experiments showed that for $E/p \leqslant 115$ V cm^{-1} mmHg^{-1} the ionization currents rise rapidly up to a gap length of about 0·5 cm and thereafter stay constant. However, it was not possible to measure these currents for the parameter

$$E/p = 115 \text{ and } 110 \text{ V cm}^{-1} \text{ mmHg}^{-1}$$

when $d > 2·5$ cm because they became unstable. Consideration of *Figure 8* together with a knowledge of the regions of unstable current (as indicated for example in *Figure 2*) shows that the region of reduced breakdown values of E/p is also the region of current instabilities, as would be expected.

The data shown in *Figure 8* were obtained by Moruzzi in this laboratory and accurately confirmed earlier measurements obtained by one of us (M.S.B.). The effect is due, as Mr. J. W. Leake showed in this laboratory, to breakdown occurring at or near the edges of the uniform field gap, i.e. in conditions of non-uniform field. It would appear that as pd (not only d) is increased above the value at which $\eta = \alpha$, the breakdown mechanism is initiated in a restricted region near an electrode (presumably the cathode) where $\alpha = \eta$ instead of at the cathode in the central region of the gap. It is still a matter of considerable interest to explain the growth of the discharge through the region, covering the major part of the spark length, where necessarily $\eta > \alpha$, and some kind of runaway process may be involved.

The electron energy distribution in SF_6 may well be peculiar because of the very high capture cross sections at very low ($< 0·1$ eV) electron energies (Ahearn and Hannay, 1953; Marriott, 1954 and Hickam and Fox, 1956). The effect shown in *Figure 8* is under further and more detailed investigation.*

It should be emphasized that the electrodes used in this investigation (Hopwood, Peacock and Wilkes, 1956) should allow gaps of at least 4 cm to be used without field distortion.

5. Conclusions

Attachment in SF_6 in swarm conditions has been studied, and the values of α/p and η/p obtained in the E/p range of 90 to 160 V cm^{-1} mmHg^{-1} show good agreement with the earlier measurements of Geballe and Reeves (Loeb, 1955) whose work is hereby considerably extended. The predominant process of negative ion formation in SF_6, in the range of parameters used, seems to be due to the capture of slow electrons producing SF_6^- and SF_5^- (primary capture process) and the contribution due to the secondary capture processes is negligible. However, it is possible that the latter processes may become important at higher pressures.

* *Note added in proof, April* 1962. Since this paper was first submitted we have undertaken a detailed study of the effect briefly discussed above and shown in *Figure 8*. We refer to the breakdown processes occurring when $\eta > \alpha$ for the discharge as a whole. The suggested runaway process invoked above is substantiated by our later work, which we hope to publish separately. The effect is not due to spurious geometrical characteristics etc., related to chamber size.

Values of the secondary coefficient γ' are very low, i.e. of the order of 10^{-7}.
Measured values of the static breakdown potentials in SF_6 are considerably higher
than that in dry air for the corresponding values of pd thereby giving high relative
dielectric strength, for example, approximately 2·8 at 380 mmHg cm, as compared
with dry air. The experimental value of $(E/p)_{lim} = 117·5$ V cm^{-1} mmHg^{-1} shows
good agreement with the predicted value of $(E/p)_{lim} = 117$ V cm^{-1} mmHg^{-1}, i.e.
when $\alpha/p = \eta/p$, thus indicating that the static breakdown in SF_6 up to
$pd = 400$ mmHg cm is brought about by the Townsend build-up mechanism except
for the anomalous behaviour shown in *Figure 8*, which is not a uniform field effect.

Acknowledgements
One of us (M.S.B.) is indebted to the Electrical Research Association for financial
support and permission to publish this paper. We are grateful to Dr. A. N. Prasad
for many discussions and advice, notably on the cathode dependence of I_0.

Appendix
The variation of I_0 with gas pressure has previously been noticed and mentioned
(Prasad and Craggs, 1961), and is almost certainly due to cathode surface changes in
a chemically active gas. The effect, for example, was not found with H_2 or N_2.
Unless the value of I_0 is found for higher pressures to fit the 5 mmHg value (see
Figure 4) where the effect is negligible and the current/voltage curve shows a
saturation value of I_0 (see *Figure 1*, cf. *Figure 2*), the values of α/p and η/p (*Figure 5*) show a pressure variation which is not to be expected. Moreover, the variation
of apparent α/p and η/p with I_0 at a given E/p is unreasonable in that both
coefficients apparently *decrease* with increasing pressure, unless I_0 is adjusted as
indicated. The change in I_0 with pressure at higher pressures (> 5 mmHg) is
associated also with the disappearance of reasonable saturation in the current/voltage
characteristic (*Figure 2*).

References
Ahearn, A. J. and Hannay, N. B., *J. Chem. Phys.* **21**, 119, (1953).
Berg, D. and Dakin, T. W., *J. Chem. Phys.* **25**, 179, (1956).
Bhalla, M. S. and Craggs, J. D., *Proc. Phys. Soc.* **76**, 369, (1960).
Dibeler, V. H. and Mohler, F. L., *J. Res. Nat. Bur. Stand.* **40**, 25, (1948).
Geballe, R. and Reeves, M. L., *Phys. Rev.* **92**, 867, (1953).
Harrison, M. A. and Geballe, R., *Phys. Rev.* **91**, 1, (1953).
Hickam, W. M. and Fox, R. E., *J. Chem. Phys.* **25**, 642, (1956).
Hochberg, B. and Sandberg, E., *Zh. Tekh. Fiz.* **12**, 65, (1942).
Hochberg, B. and Sandberg, E., *Dokl. Akad. Nauk S.S.S.R.* **53**, 511, (1946).
Hopwood, W., Peacock, N. J. and Wilkes, A., *Proc. Roy. Soc.* **A235**, 334, (1956).
Howard, P. R., *Proc. Instn. Elect. Engrs.* **104A**, 123, (1957).
Liu, T., Moe, G. and Duncan, A. B. F., *J. Chem. Phys.* **19**, 71, (1951).

Loeb, L. B., *Basic Processes of Gaseous Electronics*, California: University Press, (1955).

McAfee, K. B., *J. Chem. Phys.* **23**, 1435, (1955).

Marriott, J., *Ph.D. Thesis*, University of Liverpool, (1954).

Marriott, J. and Craggs, J. D., *Brit. J. Electronics*, [IV], **1**, 405, (1956).

Nostrand, E. D. and Duncan, A. B. F., *J. Amer. Chem. Soc.* **76**, 3377, (1954).

Prasad, A. N., *Proc. Phys. Soc.* **74**, 33, (1959).

Prasad, A. N., *Ph.D. Thesis*, University of Liverpool, (1960).

Prasad, A. N. and Craggs, J. D., *Proc. Phys. Soc.* **77**, 385, (1961).

Schumb, W. C., Trump, J. G. and Priest, G. L., *Industr. Engng. Chem.* **41**, 1348, (1949).

Introduction to Paper 14 (Haydon, 'Spark channels')

The papers 1 to 13 have largely been concerned with the earlier stages of the breakdown process, when the current densities involved are comparatively low, perhaps up to 10^{-5} A/cm^2. There has also been for many years considerable interest in the latter stages of breakdown when the breakdown process has led to the establishment of a highly conducting spark channel. However, it has only been in the last ten years or so that experimental techniques have been developed which enable information to be obtained in the intermediate time interval, close to what Haydon calls 'the birth of the spark channel'. Paper 15, which was a review paper presented at the 1967 biennial conference on ionized gases, is given here because it gives an excellent summary both of typical experimental techniques and of the main features of the data obtained. To provide the same information by giving all the original papers cited by Haydon would obviously occupy too large a part of the present volume. The theoretical papers of Köhrmann and Braginskii referred to in paper 14 are however given here as papers 15 and 16 since they give good descriptions of the early (Köhrmann) and late (Braginskii) development of the spark channel.

Further Reading
Craggs, J. D., *Spark Channels* in *Electrical Breakdown of Gases* ed. J. M. Meek and
    J. D. Craggs (Oxford University Press) 2nd Edition (to be published).

14

Spark channels

S. C. HAYDON*

1. Introduction

The last decade has seen a rapid development of new techniques for studying the
enormously complex phenomena associated with the development of spark and
other high pressure discharges. Image converters, image intensifiers and highly
time-resolved spectroscopic techniques using fast rise-time photomultipliers have
now overcome many of the limitations of the earlier experiments of Somerville and
others in Australia,[1-3] Abramson, Gegechkori, Vanyukov and others in Russia[4-5]
and Craggs and his co-workers in England.[6] These limitations restricted many
investigations to studies of events immediately following the establishment of the
highly conducting channel, at which stage a shock front develops and the phenomena
are largely determined by hydrodynamical considerations. Theories of the spark
channel behaviour in these later stages were developed by Drabkina[7] and by
Braginskii[8] and their validity tested by observations of the radius of the expanding
spark channel as a function of time. When the energy input to the channel was
effectively instantaneous, as in the case of the cable discharges used at Armidale
by Somerville and his group[3, 10] the variation of the channel radius r with $t^{\frac{1}{2}}$ predicted by Drabkina was observed. However, Drabkina's prediction that when the
shock front separates from the spark channel at later times the radius r should
vary as $t^{0.376}$ was not confirmed. Sigmond,[37] working with nanosecond arcs in
hydrogen also found the theory to be invalid. In Braginskii's theory an attempt
was made to relate the channel radius to the current $I(t)$ at time t and to make
allowance for loss of energy from the channel either by thermal conduction or
radiation. Andreev, Vanyukov and Kotolov[35] experimenting with an oscillating
spark current in air tested the predictions and found substantial agreement with
the Braginskii theory when thermal conduction losses were neglected.

Despite these encouraging successes, there persisted for some years an almost
total ignorance of the events immediately preceding the birth of the spark channel.

* Reproduced from 'A survey of phenomena in ionized gases', from the Eighth International
Conference on Phenomena in Ionized Gases. Vienna, (1967) by permission of The International
Atomic Energy Agency, Vienna.

Fortunately the degree of sophistication in the recent experimental techniques has reached a stage where considerable fine structure in the various transition stages to the highly conducting channel can now be observed. Whilst the theoretical understanding of the complex phenomena has not altogether kept pace with the experiments, nevertheless this seems to be an appropriate time to review the salient features observed in recent investigations of the earliest stages of the development of spark channels. We begin with a very brief account of the important highly time-resolved techniques before considering the various stages in the development of spark channels formed under both low and high overvoltage conditions. If liberal reference is made to the recent research efforts at Armidale, it will merely reflect the importance of the pioneering efforts of the late Professor J. M. Somerville in this field over the past 15–20 years and this review will serve as a small, though inadequate, tribute to him.

2. Methods of initiating spark channels

In recent years two different techniques have been used to apply the necessary overvoltages to the spark gaps. Since the magnitude of the percentage overvoltage applied is extremely important in determining the nature of the subsequent phenomena in the ionized channel we describe briefly the two experimental arrangements most commonly used.

(a) 'Condenser' discharges

The principle of this method is illustrated in *Figure 1*, which shows the arrangement used in recent measurements by Tholl.[12]

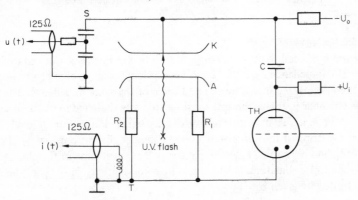

Figure 1. Circuit illustrating method used for studying pulsed 'condenser' discharges at high overvoltage. TH: Thyratron; S: Voltage divider: $R_1 = 1\,\Omega$, $R_2 = 10\,\Omega$; T: Current transformer.

Condenser C is charged to potential U_1 and the thyratron TH triggered shortly (≈ 100 ns) before electrons are released from the cathode K. The switching of the thyratron applies the additional voltage pulse U_1 to the permanently applied voltage U_0 so that $U_1 + U_0$ then exceeds the threshold voltage U_s and so overvolts the spark gap. In this way overvoltages up to ~ 50%, sufficient to create the critically high avalanche amplifications within the 2 cm gap, are applied. Typical voltage and current oscillograms obtained with this method are shown in *Figure 2*. E/p values at the threshold of breakdown in H_2 are typically ~ 25 V/cm torr.

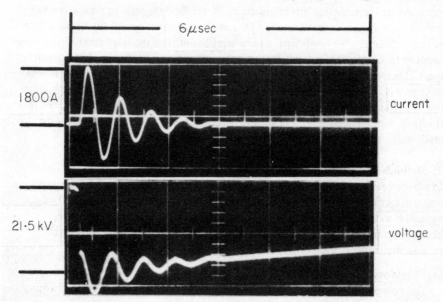

Figure 2. Current and voltage as a function of time following breakdown of the gap.

(b) Coaxial-cable discharges

These are produced with the arrangement shown in *Figure 3*. Coaxial cables of characteristic impedance Z_0 = 50 Ω and various lengths are charged to potential U_0 and then discharged into a terminated cable of the same impedance. The coaxial discharge chamber in this technique must be specially designed in order to maintain the characteristic impedance throughout the discharge circuit. For an ideal operation the firing of the gap produces a rectangular current pulse of height $I_0 = U_0/2Z_0$ and duration $2T$ where T is the pulse transit time of the charged line. Owing to the finite time required to attain high conductivity in the gap, the current rises at a finite rate given by

$$I(t) = U_0/(2Z_0 + Z(t)) \tag{1}$$

where $Z(t)$ is the discharge resistance at time t. For a pulse starting at $t = 0$, equation (1) holds to a sufficiently good approximation for most applications for times $t \leqslant 2T$.

The main advantage of this system is that from the accurately recorded current oscillogram it is possible to calculate the other important electrical quantities such as voltage, resistance and energy input into the channel.

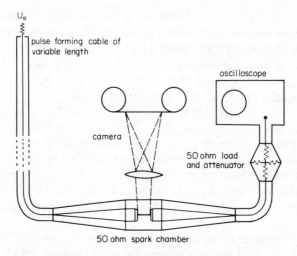

Figure 3. Experimental arrangement for studying arrested discharges with co-axial cable at low overvoltages.

Measurements with this technique have been confined so far to spark gaps of small dimensions ($\sim$ mm) with $(U_0 \text{ max}) \lesssim 10$ kV and small ($\approx 1\%$) overvoltages corresponding to values of E/p in hydrogen at the threshold for breakdown ≈ 28 V/cm torr.

No systematic investigation has yet been carried out of the effects of variations in the external circuit conditions. At the low currents observed in the earliest stages these effects are unlikely to be very significant but they may become very important as the current develops into the high ampere range at the later times. In the work described in this review each group has maintained the same external conditions for all observations with their particular technique.

3. Methods of investigating channels

The various time-resolving techniques used in spark studies may be divided into the following categories.

(a) Arrested discharges

Here the aim is to expose a photographic film to the light emitted from the spark gap during the growth of the current and to stop this flow of current at some pre-determined time. In this way a picture of the spatial distribution of light emission in the gap is obtained integrated over the full discharge duration. Time resolution is achieved by varying the arresting time. These techniques have been used by Tholl and others at Hamburg and by Schröder and others at Armidale.

(b) High-speed shutter photography

The introduction of a controlled shutter provides certain advantages over the arrested discharge method and may be achieved by mechanical or electro-optical means. Among the most convenient and rapidly operating shutters for spark studies

are the electro-optical or 'Kerr-Cell' shutters used by Dunnington[13] and by
White.[14] Their use is restricted, however, to the later phases of the channel development where the intensity of light output is high. More recently electronic tubes of the
image converter and image intensifier type have been used by Saxe,[15] Tholl,[16]
Schröder,[17] Wagner,[19] Doran and Meyer[18] and others. These devices help with
spark channel studies by providing both an overall sensitivity greater than that of
an unaided photographic emulsion and a means of fast shuttering. By operating as
camera shutters 'integrated' photographs of events occurring up to any specified
time may be obtained. An apparatus using an RCA 4449 image converter tube is
shown in *Figure 4.* Here closure of the shutter was achieved by applying to grid G1
a negative pulse of 1 μs duration and 1 ns rise-time. In order to synchronize both
discharge and shutter action, the main discharge was initiated by a flash of light
from the trigger gap TG1, and required slightly overvolted ($\sim$ 1%) conditions in
the main gap.

Some of the very early events are inaccessible to image converters because of
the lack of sufficient light intensity to meet the threshold conditions for photographic recording. Image intensifier tubes can then be used which amplify the input
light intensity by a factor of 10^5. The limiting factor for recording low-intensity
phenomena is then set by the dark-current scintillations.

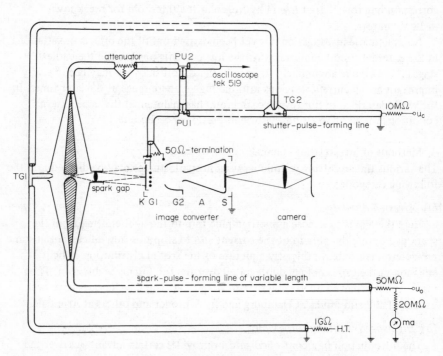

Figure 4. Apparatus for high speed shutter photography using
image converter.

(c) *Streak photography*

For adequate resolution of the fastest processes occurring in high pressure discharges, streak camera techniques must be employed, and the image intensifiers now available provide sufficient sensitivity for electro-optical shutters to be used in this way. Considerable progress has been made recently by Wagner[19] and Koppitz[11] at Hamburg and by Meyer and Cavenor[20] at Armidale. Deflecting electrodes incorporated into the image tubes permit a time sweep to be introduced into the display on the image tube screen. Usually a sharply rising voltage is applied to one pair of deflecting plates to produce a linear sweep of the image across the tube screen. An experimental arrangement used in very recent investigations by Meyer and Cavenor is illustrated in *Figure 5*. By opening the shutter of the camera for only 1/50 s, undue exposure of the film by dark current scintillations of the phosphor was avoided. Release of the shutter produced, with the help of a thyratron pulser, a trigger spark which caused the gap TG1 to break down. Ultra-violet light from this gap entered the spark chamber through a quartz window and triggered the main gap. Then the trigger gap pulse passed through a variable length of decay cable to break down TG2, which triggered the sweep generator providing a push-pull ramp voltage for the deflection plates of the converter. The pulse continued through a second delay cable to initiate breakdown in TG3. This provided a negative voltage pulse of subnanosecond risetime and microsecond duration for gating the converter

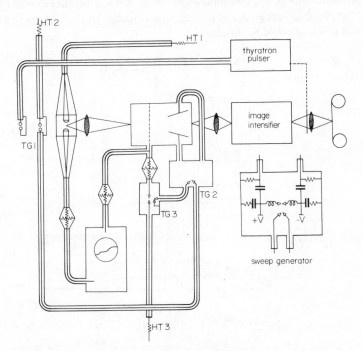

Figure 5. Apparatus for streak and shutter photography using image converter and intensifier.

tube. After closing the converter the same pulse passed on to the intensity control grid of a Tetronix Tek 519 CRO thus dimming the trace of the current pulse. In this way the development of the spark observed optically could be correlated with the increase in current oscillographically.

(*d*) *Spectroscopic investigations*

If spectroscopic studies of sparks are to yield significant information they must be capable of very high time-resolution. Very fast risetime photomultipliers are therefore essential to record the spectrally resolved intensity output of the mono- chromator. Often the studies are complicated by the varying opacities of different spark channels or the varying emissions from different atomic or molecular sources in gas mixtures such as air. Consequently considerable effort has been devoted recently to studies in relatively pure hydrogen and much of our improved under- standing of the channel formation processes is due to complementary studies in this gas carried out at Hamburg and at Armidale. The principle of the methods used is illustrated in *Figure 6* which shows the arrangement used by Meyer to study the emissions from the coaxial cable discharges produced at Armidale. Light from the required region of the spark was isolated by the entrance slit of the mono- chromator and detected by fast risetime photomultiplier. Recordings of the spectrum were made from 2500 to 5000 Å at each position and from accurate measurements of line and continuum intensities and line profiles information about the electron temperatures and electron densities respectively as functions of time and position could be obtained.

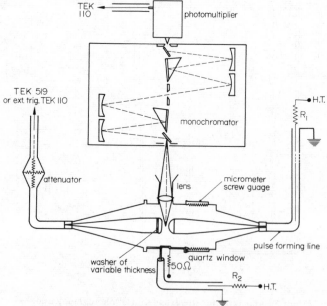

Figure 6. Arrangement used for spectroscopic investigations of hydrogen discharges at low overvoltage.

4. Experimental investigations

The sequence of events involved in the development of spark channels depends very greatly on the percent overvoltage applied to the spark gap. Marked visual differences exist between the various phases of development in the low ($\approx 1\%$) and highly (15–50%) overvolted gaps. These differences stem from the fact that at low percent overvoltage both primary and secondary avalanches contribute to the growth of current, whereas at very high overvoltages high amplifications can be produced in a single avalanche which then change the luminous appearance of the spark channel as it develops. We examine first the results of investigations with low percent overvoltage.

(a) Studies at low percent overvoltage

With low overvoltages the early stages of current growth between plane parallel electrodes may be understood in terms of the superposition of succeeding generations of avalanches. At voltages fractionally below that required to produce a spark a diffuse visible glow has been observed[21, 22] at more or less stable currents of about 1 μA. Even with moderate overvoltages dynamic studies by Schröder and by Doran[36] have shown that the current may remain diffuse to quite high current levels (≈ 100 A in N_2) before a filamentary spark channel is formed. These early stages have received a great deal of attention and theories of the formative times involved in the current growth have been propounded and tested experimentally. *Figure 7* shows a comparison between the current growth in air at atmospheric

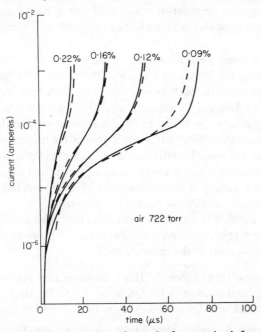

Figure 7. Comparison of growth of current in air for various overvoltages obtained experimentally (– – –) by Bandel[23] and calculated (–) by Ward.[24]

pressure measured by Bandel[23] and the theoretical curves computed by Ward[24] for a number of different overvoltages. The dominant influence of the overvoltage is clearly seen. More recently Köhrmann[25] has taken his calculations up to times where a partial voltage collapse occurs. For times greater than this, however, a satisfactory theory is not available because of the great complexity of the phenomena.

After the initial steep fall in voltage a step may occur in the voltage-time characteristic, followed by a further drop. In air at atmospheric pressure the duration of the step is usually less than a microsecond and increases rapidly as the pressure is lowered. Although the step duration is much larger in hydrogen, it may vary in random fashion from one discharge to another. Rogowski and Tamm[26] first identified the first voltage drop with the appearance of a diffuse 'glow' discharge and the second with the change to a filamentary spark channel. Saxe,[15] using an image converter tube as a fast shutter, confirmed the appearance of a bright channel developing from a diffuse discharge and from spectroscopic investigations in air showed that light from the diffuse glow consisted almost entirely of the second bands of nitrogen, with atomic lines and a continuum appearing as the channel began to develop. Subsequent studies of these phases of the discharge by Allen and Phillips[27] and by Schröder[17] confirmed the main features although Allen and Phillips did not observe any marked difference in the spectroscopy of the two phases in air and nitrogen. Schröder was able to correlate the development of the diffuse glow and the growth of luminous filaments with the current oscillograms and his observations that the rate of rise of current depends on gas pressure have also been confirmed by Vorobev, Golynskii and Mesyats.[28]

The major effort in recent years has been directed towards an understanding of the phenomena in the molecular gases H_2 and N_2. The events in H_2 can now be reproduced satisfactorily and in much more stable form. They have been studied in detail at Armidale by Meyer and Cavenor[20] who found, however, that the intensity of light output from the spark gap did not become sufficiently high to be recorded by the image intensifier until some 200 ns of the immediate post breakdown regime had elapsed. Thereafter, different phases in the development of the discharge could be identified with the structure of the current oscillogram. *Figure 8* shows a typical growth of current curve for hydrogen and helps to define the different time intervals used by the various investigators.

At time $t = t_0$ the threshold criteria for breakdown have been satisfied so that for $t < t_0$ we have pre-breakdown conditions. For times $t > t_0$ we may identify five phases in the development of the spark channel:

(1) $t_0 < t < t_a$ (see curve (a) of *Figure 8*. This is the immediate post-breakdown regime extending from currents $\approx 10^{-6}$ A up to ≈ 1 A. The time t_a normally represents the zero of time for studies in H_2 to be described in this review.

(2) $t_a < t < t_b$. The glow formation stage where the current rises to a maximum before decreasing into the next stage.

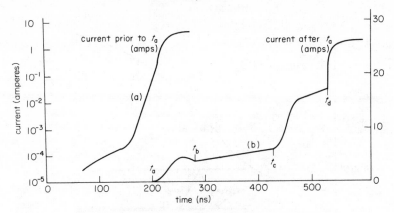

Figure 8. Typical growth of current curve for low overvoltage spark showing times to various stages of development.

(3) $t_b < t < t_c$. The diffuse glow discharge region where the current increases linearly.

(4) $t_c < t < t_d$. The filamentary glow region or the pre-channel region where a filamentary luminous channel develops out of the glow discharge column.

(5) $t_d < t$. The spark channel formation stage where the current rises extremely rapidly to establish the final spark channel which exists at later times.

We can now summarize the information that has been obtained recently about these various phases in low overvoltage spark channel development in hydrogen and nitrogen.

(i) *The immediate post-breakdown regime*

Whilst the earliest stages of this phase are, as yet, inaccessible for studies in H_2, Doran[36] has nevertheless been able to examine this important region in N_2. His results which are reported at this conference are summarized in *Figure 9* which shows the rise in current following the fourth generation of Townsend avalanche at $t = 490$ ns, a schematic drawing of streak photographs and the sequence of integrated shutter photographs. The later stages, as we shall see, resemble the general features observed in hydrogen and Doran's work in N_2 now gives some further insight into the earliest stages of the development in H_2 that have so far been inaccessible to observations. The velocity of the luminous front, initially moving towards the cathode, increases from 5×10^7 cm/s to 10^8 cm/s. Subsequent fronts moving to and fro in the gap have velocities $\approx 2 \times 10^8$ cm/s and have been shown by Doran to be associated with changes in the local space charge induced fields which locally enhance ionization as they propagate through the gap. His measurements, made at a pressure of 300 torr with a gap separation of 2 cm and a percent overvoltage of 7·56%, represent an important contribution providing additional information to supplement the extensive studies at low ($\approx 1\%$) and high (up to $\approx 50\%$) overvoltages in hydrogen which are reviewed in this paper.

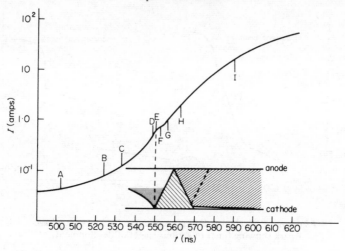

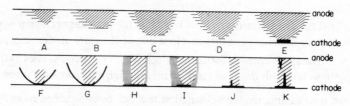

Figure 9. Growth of current, schematic drawing of streak
photographs and sequence of integrated shutter photographs
for spark channel development in N_2 (reference 36).

(ii) *The formation of a diffuse discharge in hydrogen*

Figure 10 shows a series of shutter photographs taken by Meyer and Cavenor[20] using the apparatus shown in *Figure 5*. In each photograph the gap separation was 2 mm, the position of the cathode, C, being indicated. In order to record satisfactorily the luminosity in the early stages in the cathode region the magnification was made as large as possible. This resulted in the light output from the later stages of the spark channel at distances greater than approximately 1·4 mm from the cathode being beyond the field of view of the converter photo-cathode. The shutter closure time of each spark is indicated by a mark on the graph of the rise in current so that the first four frames record the development of the discharge during the rise in current immediately preceding and following the time t_a. These same frames indicate a diffuse luminous region developing in front of the cathode, extending progressively towards the anode. At the time t_4 when the current reaches a maximum the luminous region extends throughout the whole of the electrode space but appears considerably more intense in the region just before the cathode. The development of the discharge radius in these times follows the current pulse as seen by the streak photograph in *Figure 11* obtained by Meyer and Cavenor.

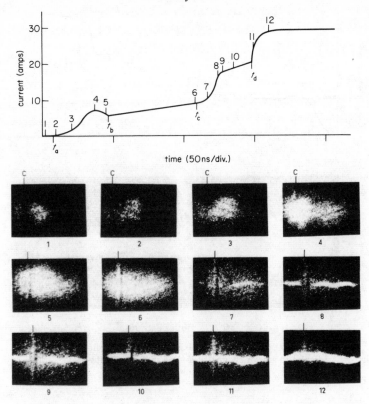

Figure 10. Series of shutter photographs showing development
of spark channel in hydrogen. Shutter closure times are
indicated on current oscillogram. [p = 500 torr; d = 2 mm;
C indicates position of cathode.]

Here a region 0·4 mm in front of the cathode was singled out by a narrow slit and
the diameter of the discharge in 650 torr of H_2 was observed with the time axis
parallel to the channel. The radius increases rapidly to a value of about 0·5 mm
before decaying slightly to attain a minimum after about 20 ns. Meyer's spectroscopic studies[29] of the thin glow in front of the cathode during the first current
step revealed the spectrum shown in *Figure 12*. This continuum emission with a
maximum below 3000 Å is caused by molecular hydrogen and is usually observed
in high pressure glow discharges. No evidence of Balmer line emission was observed
in these early stages.

(iii) *The diffuse glow discharge in hydrogen*

Frames 5 and 6 of *Figure 10* were taken with the aperture of the first lens slightly
reduced and show the appearance of the discharge during the glow t_b-t_c. At time
t_b a structure is first observed consisting of a bright luminous region in front of
the cathode surface, a narrow dark space and uniform glow extending to the anode.

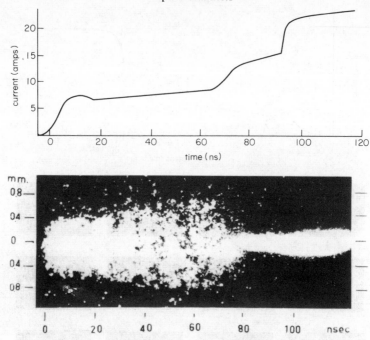

Figure 11. Streak photograph in hydrogen with slit perpendicular to axis of channel at d = 0·4 mm from cathode.
[p = 650 torr, gap separation 1 mm.]

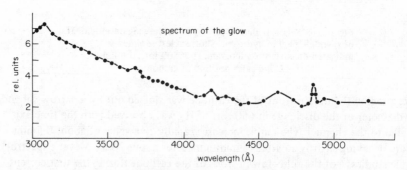

Figure 12. Spectrum of the light emission from the thin glow
in front of the cathode during the first current step.

This structure is maintained up to time t_c.

Spectroscopic observations of the glow phase recorded for a pressure of 650 torr
from a region 0·4 mm in front of the cathode reveal a continuum with maximum
intensity at λ = 3100 Å and the appearance, with low intensity, of the Balmer
H_β-emission. Observations of the variation of the intensity of the λ = 3100 Å
emission shows that this drops sharply to very low levels as the current rises into
the second step (*Figure 13*).

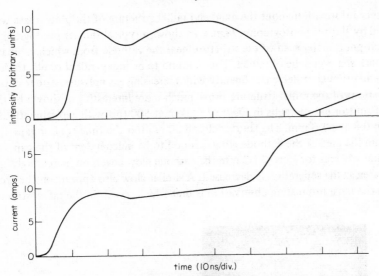

Figure 13. Variation of the intensity of the emission at
$\lambda = 3100$ Å in the diffuse glow and early filamentary glow
phases. [$p = 650$ torr, position 0·4 mm from cathode, gap
separation 1 mm.]

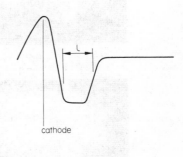

Figure 14. Typical shutter photograph and idealized microphotometer scan of the luminous structure near the cathode
during the diffuse glow stage.

Further information about the axial and radial structure of the glow phase was obtained by shutter photography. *Figure 14* shows a typical shutter photograph and microphotometer scan of the structure near the cathode from which estimates of the dark space may be obtained. This is found to be independent of electrode separation although it decreases linearly with increasing gas pressure. The results of investigations of the radial structure prove much more interesting as shown by the series of shutter photographs in *Figure 15* taken at 500 torr with gap separations between 0·8 and 1·8 mm. The shutter closed 20 ns after the time t_b. For gaps > 1·2 mm the area of the cathode glow is found to be independent of the gap separation whereas for gaps < 1·2 mm the cathode glow covers an increasingly greater area as the separation is decreased. A similar glow also appears on the anode. Perhaps the most interesting observation is that which reveals in the radial structure

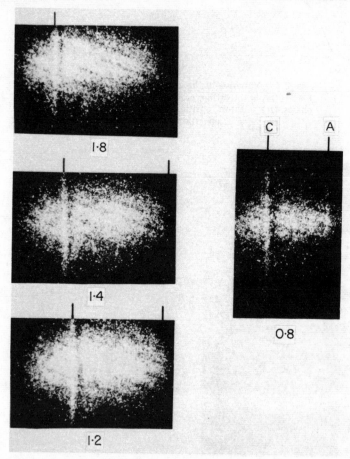

Figure 15. Series of shutter photographs taken at p = 500 torr in hydrogen with gap separations between 0·8 and 1·8 mm, showing the similarity in the luminous structure for $d > 1·2$ mm.

a central peak in intensity, surrounded by a less intense luminous ring. At the later stages of this glow discharge phase the central peak becomes increasingly intense and more concentrated at the axis.

(iv) *The development and growth of the luminous filament*

Frames 7–10 in the series of shutter photographs of *Figure 10* show an increase in luminous intensity occurring along a thin filament. These filaments originate near the anode and then extend across the gap towards the cathode. When the filament reaches the dark space a second quasistable state is attained and the rate of growth of current slows down. These phenomena occur in the time interval $t_c < t < t_d$. Measurements of the discharge current show that at constant electrode separation the current increases with increasing gas pressure. Furthermore, the voltage across the gap immediately prior to the transition to the arc is nearly the same for all gap separations $d > d_{crit}$. At $d < d_{crit}$ this voltage is lower and when d is very small no transition from glow to arc or glow to spark takes place. During the rise of current into the second step the intensity at 3100 Å falls to some lower level whereas the Balmer lines H_α, H_β and H_γ increase in intensity, as does the

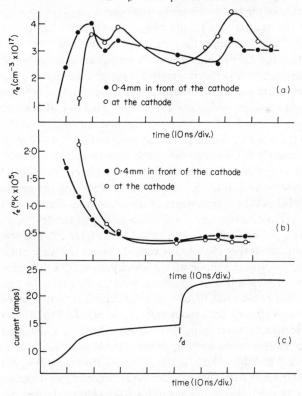

Figure 16. Variations of electron density, n_e, and electron temperature T_e at two positions in the gap during the filamentary glow and final spark channel phases.

continuum, above $\lambda \approx 4000$ Å. The profiles of the H_β and H_γ lines measured at $p = 650$ torr have been compared with Griem's theoretical profiles and in this way Meyer has established the variation of ion density n_i with time shown in *Figure 16(a)*. Further measurements of the ratio of H_β-line/continuum yielded the variation of T_e with time shown in *Figure 16(b)*. With these values and the theory for continuum emission of H-atoms the relative continuum intensity in the near UV has been calculated and compared with the observed spectra. In this way it was possible to show that at the time when the current reaches the second step all radiation below 3600 Å is caused by bremsstrahlung and recombination of atomic hydrogen. No molecular emission appears at this stage.

The streak photographs (*Figure 11*) show clearly the decrease in radius of the glow discharge phase. The intensity of the luminous ring decreases whilst the central peak gains in luminosity until with the rise into the second step of the current oscillogram the discharge shows a well-defined filamentary appearance. Throughout this period the cathode dark space remains unbridged by the developing filamentary channel and no evidence of a cathode spot is obtained.

(v) *The formation of the spark channel in hydrogen*

At t_d, some 150 ns after time t_a, the shutter photograph 11 (*Figure 10*) reveals that a thin bright junction bridges the dark space, at the same time that the current rises suddenly into the state where a complete voltage collapse occurs. With the increase in luminous intensity this integrating method must rely on a change of aperture in order to record the event satisfactorily and it is only with streak photography (*Figure 11*) that any ambiguity about the significance of the apparent disappearance of the broad diffuse luminous region is removed. With the final rise of current into the phase of complete voltage collapse the intensity of the spark channel increases and a radial expansion of the filament can be detected in frame 12 (*Figure 10*). The rise of current in the final step is very rapid increasing by some 5 A in 1 ns, is independent of the gap distance but varies with the gas pressure.

Further detail about the crucial stages of transition from glow-to-arc were obtained by Meyer and Cavenor by streak photography of the developing channel with the time axis perpendicular to the channel as in *Figure 17*. This reveals the formation of a luminous spot close to the cathode before the wave of luminous excitation moves towards the anode with a velocity, $v_a = 2 \cdot 2 \times 10^6$ cm/s. One also notes from these photographs a much slower 'wave' moving into the gap having started at the anode at the time, t_d, at which the current rise into the second step occurs. About 30 ns after t_d both waves meet in the gap close to the anode.

Spectroscopic investigations again led to values of both n_e and T_e and when the voltages starts to collapse completely n_e at the cathode rises rapidly to a maximum of $4 \cdot 4 \times 10^{17}$ cm^{-3} and then decays again. Some few nanoseconds later this increase is noted 0·4 mm from the cathode. No evidence of metal vapour lines was obtained until very late in the development (≈ 300 ns after t_d) when the current ceases. At this stage zinc lines associated with the metal vapour could then be observed.

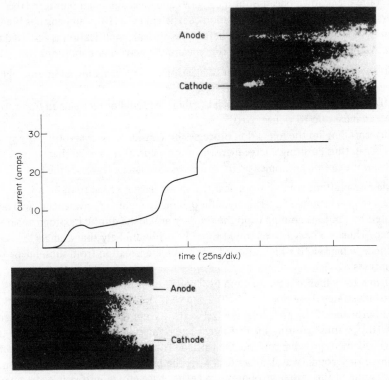

Figure 17. Streak photographs of glow-to-spark transition taken
with time axis perpendicular to the channel. Positions of
photographs correlate with time axis of current oscillogram.

(b) Studies at high percent overvoltages

With sufficiently high percent overvoltage it becomes possible for a single
avalanche to create conditions in mid-gap which favour the development of
relatively narrow luminous filaments, referred to as streamers. We first consider
briefly the conditions required for their appearance and subsequent growth into a
spark channel.

(i) *Channel development from a single avalanche*

The significant parameter is generally believed to be the total number n, of
electrons (or positive ions) in the avalanche. Depending on the gas, overvoltage and
other conditions the critical value of n is usually taken as 10^8 to 10^9 electrons so
that n_0 initial electrons traversing a distance x will have a good probability of
initiating an avalanche-streamer-spark channel sequence of events if n [$n_0 \exp (\alpha x)$]
becomes $= 10^9$. These critical amplifications must be attained at distances $x < d$
(the gap separation) and cannot be achieved in a single avalanche under conditions

near static breakdown because for most gases and electrode materials ω/α (the generalized second Townsend ionization coefficient) is $\geqslant 10^{-9}$ and ensures breakdown by the Townsend mechanism before the critical amplification is achieved at $x < d$ (reference 29). One can, however, ensure the necessary conditions by:

(i) increasing the initial number of electrons n_0 by focussing an intense light flash on to a spot on the cathode;

(ii) adding an organic vapour such as ether, methalyl or methane in order to reduce ω/α to values $< 10^{-9}$, or

(iii) speeding up the ionization processes by considerably overvolting the spark gap thus ensuring a large increase in the value of α to α' so that $n \times \exp(\alpha'd)$ becomes $= 10^9$ before the avalanche traverses the gap.

In this case photons coming from near the anode have no time to build up successive generations of avalanches although many secondary avalanches will be initiated by photons coming from the primary avalanche during its progress across the gap. Thus in a heavily overvolted gap it is highly unlikely that a single avalanche will grow to a high value of n in complete isolation from subsequent generations.

Figure 18, which is a photograph by Tholl[30] of a suppressed overvolted discharge, illustrates these points well. The gap was 3 cm wide in N_2 at 300 torr, the time of crossing of the avalanche was $2 \cdot 1 \times 10^{-7}$ s and the voltage removed after $3 \cdot 4 \times 10^{-7}$ s, thus arresting the discharge. The primary avalanche has given rise to a bright filament growing towards the cathode from the anode and the light from the cloud of secondary avalanches following the primary one can be clearly seen. In fact some of the earlier secondary avalanches have grown into visible streamers under the influence of the space-charge distorted field and may be seen as streaks near the anode.

Figure 18. Photograph of suppressed overvoltage discharge in N_2 at 300 torr (reference 30).

(ii) *The anode- and cathode-directed streamers*

When an electron avalanche has attained the critical carrier amplification in the gap, anode- and cathode-directed streamers then appear. Allen and Phillips[27] observed a narrow bright region in the cathode streamer behind the point where these critical conditions were believed to be satisfied and Tholl[31] noted its persistence as a constriction or 'neck' in the latest stages of the spark channel development when the current exceeded 10 A. Wagner[32] also examined these streamers and estimated the velocity with which the luminosity extends in each direction. In a coordinate system moving with the drift velocity of the electrons he found the velocities to be the same and suggested that it is the streamer electrons that are responsible for further propagation of the streamer. At a later stage in the development Wagner also observes an abrupt intensity increase and accelerated propagation of the cathode-directed streamer. He attributes this to the effect of secondary electron avalanches merging with the cathode-directed streamer having been initiated at the cathode by a photoelectric effect. The final bright spark follows the arrival of the cathode-directed streamer at the cathode when a series of very fast luminous fronts propagate across the gap as observed by Koppitz.[11]

Tholl[31] has also investigated these streamers using both arrested discharge and spectroscopic techniques. He found, for hydrogen, that, with increasing percent overvoltage, the constriction appeared at smaller and smaller distances from the cathode as shown in *Figure 19*. This figure also shows the degree of correlation between the distance x_k at which $n_0 \exp (\alpha x_k) = \exp (20)$ and the observations.

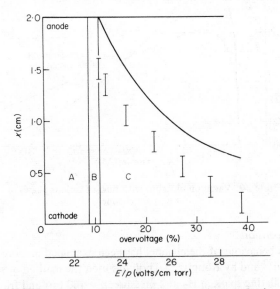

Figure 19. Position of constriction (or 'neck') appearing in hydrogen as a function of percentage overvoltage. Full line is obtained from criterion, $n_0 \exp (\alpha x_k) = \exp (20)$; $n_0 = 330$.

(iii) *The formation of the spark channel*

By extending his studies to hydrogen Tholl was also able to record the distribution of temperature and ion density in the developing channel. In the constriction the ion density was lower than in other parts of the spark. Outside the constriction the density is at first higher but then rapidly decreases to values lower than that in the constriction. Also the temperature in the constriction decreases more slowly than elsewhere and within 500 ns temperatures decrease from about $2 \times 10^{4\circ}$K to about $8 \times 10^{3\circ}$K. The pre-discharge channel established by the anode- and cathode-directed streamers has a relatively low conductivity $\approx 10^{-3}$ A/V cm but as more energy is delivered to it a plasma channel with much higher conductivity $> 10^{-1}$ A/V cm is formed. The axial distribution of temperature and ion density is very inhomogeneous in this plasma channel and according to Koppitz[11] this is responsible for the axial inhomogeneities observed in his recent investigations of the radial and axial development of the hot plasma spark channel of the later stages. Tholl also measured the channel radius and showed that the channel expansion at first begins at the constriction or 'neck', followed later in the anode- and cathode-directed parts. After about 50 ns, *Figure 20* shows the channel radius remaining approximately constant in the constriction but decreasing at different rates in the cathode- and anode-directed streamers. The maximum radius of the channel (≈ 0.7 mm) occurred in the anode-directed streamer. In the cathode part and in the constriction, the radius did not exceed 0.5 mm.

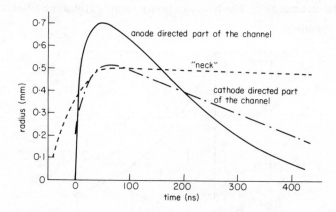

Figure 20. Variation of radius with time of various phases of the spark channel.

(iv) *Channel expansion*

This complex behaviour of the channel has been examined in more detail recently by Tholl[12] and by Koppitz.[11] Tholl has used the method of section 11(a) to obtain an oscillating spark in H_2 at a pressure, $p = 460$ torr and from spectroscopic measurements found the temporal variation of the radial distribution of electron density and temperature. At the maximum current in the first half period

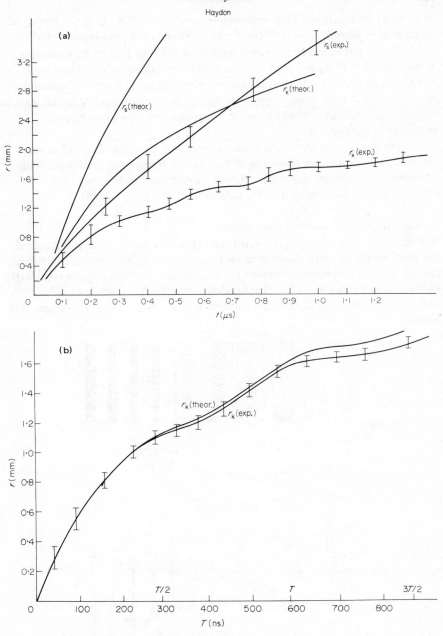

Figure 21. (*a*) Comparison with the theory of Drabkina[7]:
$C = 10\text{pF}; L = 0\cdot84\ \mu\text{H}; p = 460$ torr;
$U_0 = 26\cdot5$ kV.
(*b*) Comparison with the theory of Braginskii[8]:
$C = 10$ pF; $L = 0\cdot84\ \mu\text{H}; R_d = 2\cdot8\ \Omega; I_{max} = 1140$ A;
$p_0 = 460$ torr in H_2.

(= 1800 A) the maximum axial temperature was $5 \cdot 2 \times 10^{4} {}^{\circ}$K. At the beginning of the channel expansion $n_e \approx 10^{18}$ cm^{-3} and at the end of the expansion $\approx 10^{17}$ cm^{-3}. This corresponded to a pressure increase at the beginning of the expansion of the channel $\approx 20\,p_0$ which by the end of the first half period decreased to $\approx 3\,p_0$ and agrees well with the pressure variation calculated from the expansion velocity of the channel using Braginskii's theory. The energy supplied to the spark is expended in approximately equal amounts between ionization, heating and expansion of the spark channel.

Further information about the radial expansion of the spark channel has been obtained by Koppitz in an extension of the work of Wagner. He finds that for his experimental conditions Braginskii's theory is much more satisfactory than Drabkina's as can be seen by the comparisons in *Figure 21(a)* and (*b*).

5. Interpretation of observed phenomena

The schematic diagrams of *Figure 22(a)* and (*b*) summarize the general features of the development of spark channels under conditions of (a) low percent overvoltage (b) high percent overvoltage. Detailed theories capable of explaining the complexity of the phenomena are not available at present. However it is possible to offer reasonable explanations of the main features in each case and these are summarized below.

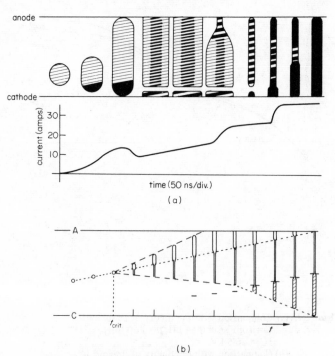

Figure 22. Schematic representations of main features of spark channel development (a) at low percentage overvoltage, (b) at high percentage overvoltage.

(a) Low overvoltage phenomena

The glow structure develops from successive avalanches in the manner treated theoretically by Köhrmann.[25] The initially uniform field becomes distorted by positive space charge near the anode as observed in N_2 by Doran and a progressively greater field then develops on the cathode side as the space charge maximum moves towards the cathode setting up a high electric field in this region. This high field gives rise to positive-ion induced electron emission in addition to photoemission and the electron current increases. Space charge increases further and the concentration of enhanced field leads to the growth of the narrow intense glow in front of the cathode. As the current continues to increase an increased field gradient develops near the anode and it is in this region that the filamentary glow appears and where the production of hydrogen atoms by dissociation of H_2 takes place. The higher conductivity leads to distortion of the field lines so increasing the field on the cathode end of the filament which consequently extends towards the dark space. With the development of this highly conducting filamentary glow the gradient across the cathode fall and dark space increases making possible the sharp transition to a more efficient cathode mechanism involving perhaps field emission. When the cathode fall conducts the increased current the final collapse of voltage takes place and for H_2 we then observe the ionizing wave move through the gap leaving behind the highly conducting channel.

(b) High overvoltage phenomena

In these conditions the appearance of luminosity in mid-gap is associated with the high electron carrier amplification in the initial avalanche. The subsequent development of luminosity in both anode and cathode directions with equal velocities as measured relative to the drifting coordinate system, suggests that this transition from avalanche to streamer is being controlled by the electrons in the initial avalanche. It is clear from the increased luminosity appearing in the cathode streamer that at some later stage the secondary avalanches produced by a surface photo effect contribute to the current growth. However it is not so obvious in view of earlier considerations of the role of photoionization in the breakdown mechanism[33] that because of the exponential dependence on E/p of the normalized velocity of the anode-directed streamer that photoionization and ionization by impact in the spacecharge enhanced field are essentially effective in the streamer propagation in pure nitrogen. A theory to explain how this can occur in pure gases has not yet been satisfactorily developed.

Once the highly conducting spark channel is formed and almost fully ionized conditions exist with electron densities $\approx 10^{18}$ cm^{-3}, excess energy given to the electrons will be distributed equally among all gas particles within a few nanoseconds because of the very high cross-sections of electron-ion interaction. This means that through ohmic heating of the ionized gas inside the spark channel the energy stored in the external circuit is converted into kinetic energy within a very short time after the channel becomes 'hot'. The pressure increases rapidly, the hot plasma acts as a piston on the surrounding gas as envisaged in Braginskii's spark model.

A shock wave is generated by the action of this piston which accelerates the shock as long as any considerable amount of energy is released inside the spark channel. For a condenser discharge this seems to last as long as the first half wave of current and Braginskii's formula for the spark channel radius holds surprisingly well indicating that heat conduction losses from the channel, not taken into account by Braginskii, may be neglected.

If the energy release ends at a time when the pressure generating the shock is very much greater than the external pressure, p_0, so that p_0 may be neglected, then the shock from then on becomes very similar to a blast wave which in the case of cylindrical symmetry gives an expansion $\alpha t^{\frac{1}{2}}$. Drabkina's formula for the shock wave holds for this case but *only* if one can neglect the time of energy release as compared to the time during which the shock expands freely. When it becomes necessary to consider the external pressure the self-similarity solutions for the shock wave no longer hold and one must resort to approximate solutions such as those of Sakurai.[34] For the cable discharges investigated at Armidale by Somerville, Fowler and others, the time for energy release is very short indeed so that the $t^{\frac{1}{2}}$ expansion law holds in the early stages of the development where the shock wave and channel boundary are not separated.

In summary this review has presented the main features that now appear to determine the growth of spark channels under widely different conditions of overvoltage. Emphasis has been given to phenomena leading up to the final stages of the spark channel formation and no attempt has been made to deal with the minutiae of the complex phenomena.

Acknowledgement
The author would like to thank G. A. Schröder, J. Meyer, A. Doran and M. Cavenor for their important contributions to the research presented. He is also indebted to Drs. H. Tholl and J. Koppitz for making their research data available prior to publication and to Dr. Meyer and Mr. Cavenor for their helpful discussion of, and preparation of material for, the manuscript.

References
1. Somerville, J. M. and Blevin, W. R., *Phys. Rev.* **76**, 982, (1949).
2. Somerville, J. M. and Grainger, C. T., *Brit. J. Appl. Phys.* **7**, 400, (1956).
3. Somerville, J. M. and Williams, J. F., *Proc. Phys. Soc.* **74**, 309, (1959).
4. Abramson, I. S. and Gegechkori, N. M., *J. Exp. Theor. Phys.* (*USSR*) **21**, 473, (1951).
5. Vanyukov, M. P., Isaenko, V. I. and Khazov, L. D., *J. Tech. Phys.* (*USSR*) **25**, 1248, (1955).
6. Craig, R. D. and Craggs, J. D., *Proc. Phys. Soc.* **B.66**, 500, (1953).
7. Drabkina, S. I., *J. Exp. Theor. Phys.* (*USSR*) **21**, 473, (1951).
8. Braginskii, S. I., *Soviet Phys. JETP* **7**, 1068, (1958).
9. Haydon, S. C., Ed. *Introduction to Discharge & Plasma Physics*, University of New England, Chapter 26, 390, (1964).

10. Creagh, D. C., Fletcher, N. H. and Somerville, J. M., *Proc. Phys. Soc.* **81**, 480, (1963).

11. Koppitz, J., *Z. Naturf.* **22a**, 1089, (1967).

12. Tholl, H., *Z. Naturf.* **22a**, 1068, (1967).

13. Dunnington, F. G., *Phys. Rev.* **38**, 1535, (1931).

14. White, H. J., *Phys. Rev.* **46**, 99, (1934).

15. Saxe, R. F., *Brit. Appl. Phys.* **7**, 336, (1956).

16. Tholl, H., *Z. Naturf.* **19a**, 346, (1964).

17. Schöder, G. A., Proc. VII Int. Conf. on Phen. in Ionized Gases, Belgrade **1**, 606, (1966).

18. Doran, A. and Meyer, J., *Brit. J. Appl. Phys.* **18**, (1967).

19. Wagner, K. H., *Z. Phys.* **189**, 465, (1966).

20. Meyer, J. and Cavenor, M., (to be published).

21. Kachikas, G. A. and Fisher, L. H., *Phys. Rev.* **88**, 878, (1952).

22. DeBitetto, D. J. and Fisher, L. H., *Phys. Rev.* **104**, 1213, (1956).

23. Bandel, H. W., *Phys. Rev.* **95**, 1117, (1954).

24. Ward, A. L., Proc. VI Int. Conf. on Ionization Phenomena in Gases, **2**, 313, (1963).

25. Köhrmann, W., *Z. Naturf.* **19** 246, 926, (1964).

26. Rogowski, W. and Tamm, R., *Arch. Elektrokal* **20**, 107, 625, (1928).

27. Allen, K. R. and Phillips, K., *Proc. Roy. Soc.* **A274**, 163, (1963).

28. Vorobev, G. A., Golynskii, A. I. and Mesyats, G. A., *Soviet Phys. Tech. Phys.* **9**, 1658, (1965).

29. Meyer, J., *Brit. J. Appl. Phys.* **18**, (1967).

30. Tholl, H., Dissertation, Universität Hamburg (1964).

31. Tholl, H., Proc. VII Int. Conf. on Phenomena in Ionized Gases, Belgrade **1**, 620, (1966).

32. Wagner, K. H., Proc. VII Int. Conf. on Phenomena in Ionized Gases, Belgrade **1**, (1966).

33. Dutton, J., Haydon, S. C. and Llewellyn Jones, F., *Proc. Roy. Soc.* **A218**, 206, (1953).

34. Sakurai, A., *J. Phys. Soc. Japan* **8**, 662, (1953); **9**, 256, (1954); **10**, 1018, (1955).

35. Andreev, S. T., Vanyukov, M. P. and Kotolov, A. B., *Soviet Phys. Tech. Phys.* **7**, 37–40, (1962).

36. Doran, A., Phenomena in Ionized Gases: Contributed Papers (Proc. Conf. Vienna 1967) Springer-Verlag, Vienna, 199, (1967).

37. Sigmond, R. S., Proc. VII Int. Conf. on Phenomena in Ionized Gases, Belgrade **1**, 611, (1966); *Proc. Phys. Soc.* **85**, 1269, (1965).

Introduction to Paper 15 (Köhrmann, 'The development of the Townsend discharge, with time, to breakdown')

The paper by Köhrmann is aimed at calculating the temporal growth of a Townsend discharge up to the time at which the collapse of the discharge voltage occurs. The calculations are carried out for currents well above those previously considered. Four stages in the development of the discharge are distinguished and their behaviour examined. The theoretical treatment is, as discussed by Haydon in paper 14, in general agreement with experiment. As successive avalanches develop in the discharge gap, the applied electric field becomes distorted by the positive space charge created near the anode. The space-charge maximum moves towards the cathode, thus increasing the electric field near the cathode. This field encourages secondary electron emission from the cathode by positive-ion bombardment in addition to that by photons, and the current increases. As the process continues a narrow intense glow ('cathode fall') first develops in front of the cathode, followed by the development of a high-gradient field at the anode. A filamentary glow appears in this latter region, the field lines become distorted, and the filamentary glow extends towards the cathode. The voltage gradient across the cathode-fall region and the dark space in front of it increases considerably and so does the secondary emission from the cathode. Finally, the voltage collapses and both from Köhrmann's work and from experimental studies an 'ionising wave' is observed to travel from anode to cathode leaving behind it a highly conducting channel. The subsequent development of the spark channel is described in paper 16.

Further Reading
Miyoshi, Y., *Phys. Rev.* **103**, 1609, (1956) and **117**, 355, (1960).
Davies, A. J., Evans, C. J. and Llewellyn Jones, F., *Proc. Roy. Soc.* **A281**, 164, (1964).
Also references given in papers 14 and 16.

15

Development of the Townsend discharge, with time, to breakdown

W. KÖHRMANN*

The current rise of the Townsend discharge was calculated for a plane discharge gap in hydrogen (pd = 1000 torr cm); the calculation starts from the original development with no space charge, and is taken over the stage where the space charge of the positive ions and electrons is active to the point where the voltage across the discharge gap falls. The model used for the calculation includes the ordinary processes like carrier movement and collision ionization (represented by the Townsend differential equation) along with photoelectric production at the cathode (γ_{Ph} effect). An essential factor for these processes is the influence of the field strength which, according to the Poisson equation, is related to the resulting space charge of the ions and electrons. The results of the calculations enable the following stages in the development of the discharge to be distinguished: during the initial build-up of the space charge (A) the development of the current is accelerated. In the next stage (B) a cathode fall is formed which supplies the electrons for the remainder of the zone. A further rise in current in this region is produced by an ionization wave (C): this wave passes (at about five times the electron drift velocity) from the anode in the direction of the cathode. In the final stage (D) the discharge consists of the cathode fall and an associated plasma in which there is no longer any noteworthy ionization. The restricted conductivity of this plasma, along with the resistance in circuit with the discharge gap, means that the voltage is only partially broken down at the discharge gap. This partial break-down was observed during earlier experimental studies, and has been taken as a characteristic of the Townsend structure. The results of the calculation are thus confirmed experimentally.

The progressive development of the Townsend discharge into a high-current type of static discharge can be explained in terms of its current voltage properties, which show a continuous transition from the Townsend discharge to the spark discharge. A static measurement of the current-voltage characteristic comes up

* Reproduced from *Zeitschrift für Naturforshung*, **19a**, 926, (1964).

against difficulties, particularly at high pd, because the discharge is to a high degree unstable. Thus it is preferable to use methods in which the transition to the high-current type of discharge is followed continuously.

Experimental studies of this kind (in the homogeneous field of a discharge gap) have been carried out in various gases at low current densities.[1,2] The increases in current observed can be explained by a simple theory, based essentially on the assumption of a modified gas amplification, $\int \alpha \, dx$, of the electron avalanches in the space-charge field of the positive ions. For this theory to be valid the field distortion must be low (about 10% at maximum) so that the customary picture of the course of the discharge with time can be used.[2-4] However, these increases in current, which amount to about 10^{-5}-10^{-4} A/cm^2, are only the start of the process that leads to breakdown.

An essential later stage in the Townsend discharge can be seen from observations of the course of the voltage at the discharge gap. Studies in air, hydrogen and nitrogen[5,6] have shown that the voltage only undergoes partial breakdown, to a value of $(0·6 \ldots 0·8)U_D$. The associated current density of about 10^{-1} A/cm^2 is about three orders of magnitude higher than in the current increases mentioned above. At the partial breakdown stage the discharge occupies the whole cross-section of the discharge gap. The spark does not appear until there has been complete breakdown of the voltage.[7] Since (under the same experimental conditions) complete voltage breakdown can occur through the kanal mechanism, the voltage oscillogram can be used as an empirical criterion for the boundary between the Townsend mechanism and kanal mechanism.[5,6]

The aim of the present paper is to calculate the rise in current with time of the Townsend discharge above the present limit of 10^{-4} A/cm^2 up to the voltage collapse region. The model used for the calculation is one dimensional with the current density uniformly distributed across the discharge gap. This means that only x and t appear as independent variables. It is also assumed that only the usual processes such as collision ionization, charge-carrier movement and feedback to the cathode due to the photoeffect are active. The effect of the field strength (which according to the Poisson equation is related to the charge distribution) is decisive for these processes. The assumptions made about these processes are set out in full in section 1. In line with the experimental conditions in reference 6, we assume that the discharge gap (with a self-capacitance C) is connected to the current supply by a series resistance R.

The calculation is carried out for a special example (as regards type of gas, pressure, distance, overvoltage etc.). Detailed discussion shows that the following discharge parameters can be chosen at will: $\alpha_0 d$, v_+/v_-, and the course of $\alpha/\alpha_0 = f(E/E_0)$ (where E_0 and α_0 relate to the onset of the discharge). The over-voltage (or the value μ_{ph}) seems to have a great effect on the course of the discharge. However, the theory valid for low current densities shows[4] that rises in current which have initially different courses take on a uniform (asymptotic) course under the influence of the space charge. The overvoltage thus seems to be a free discharge parameter. Of the other discharge parameters, only α/α_0 can basically be varied, and that by the selection of another value for pd. If we limit ourselves to $pd > 100$ torr cm, the example given here is of general validity.

A paper published recently calculates the rise in voltage for a Townsend discharge (air, $pd = 760 \cdot 1$ torr cm),[8, 9] but on the basis of a uniform voltage in the discharge gap. Thus it is not possible to compare these results directly with the ones given here.

1. Basic equations

The production of electrons and ions in a gas space is described by the Townsend differential equation (for its generalized form, see reference 10):

$$\frac{\partial n_-}{\partial t} = -\frac{\partial j_-}{\partial x} + \alpha j_-, \quad j_- = v_- n_- \tag{1}$$

$$\frac{\partial n_+}{\partial t} = \frac{\partial j_+}{\partial x} + \alpha j_-, \quad j_+ = v_+ n_+ \tag{2}$$

The development of the discharge over a period of time is determined by the electron current leaving the cathode (equation (3)). Equation (4) is a boundary condition for the ion current:

$$j_-(0, t) = j_F(t) + \int_0^d j_-(x, t)\, \delta(x, t)\, dx$$

$$= j_F(t) + \gamma_{ph} \int_0^d j_-(x, t)\, \alpha(x, t)\, dx \tag{3}$$

$$j_+(d, t) = 0 \tag{4}$$

The field conditions in the discharge gap can be obtained from the Poisson equation:

$$\epsilon_0 \frac{\partial E}{\partial x} = n_- - n_+ \tag{5}$$

The discharge gap (with self-capacitance C) is connected to the current source U_0 by the resistance R (*Figure 7*). The following then applies:

$$\frac{1}{R}\left\{U_0 - U(t)\right\} = C\frac{dU(t)}{dt} + I(t) \tag{6}$$

with
$$U(t) = \int_0^d E(x, t)\, dx$$

$$I(t) = F\{J_+(t) + J_-(t)\} \tag{7}$$

$$J_+(t) = \frac{1}{d}\int_0^d j_+(x, t)\, dx \tag{8}$$

$$J_-(t) = \frac{1}{d}\int_0^d j_-(x, t)\, dx \tag{9}$$

F = electrode surface or discharge cross-section.

Equations (6) to (9) have been used hitherto only for the movements of a few charge carriers in a practically homogeneous field; a derivation of these expressions which also proves their applicability to strongly distorted fields will be found in reference 9.

The drift velocities v_- and v_+ are given as linear functions of E/p. Various analytical expressions are used for the collision ionization coefficient α depending on the E/p region. The coefficient δ is assumed to have the same dependence on field strength as for α: $\delta = \gamma_{ph}\alpha$.

The example treated has the following data:

Hydrogen: $p = 500$ torr, $d = 2$ cm.

$$\gamma_{ph} = 8\cdot34 \times 10^{-4}; E_0 = 1\cdot984 \times 10^4 \text{ V cm}^{-1}:$$

$$(U_0 - U_d)/U_d = 0\cdot2\%; \ \mu_{ph} = 1\cdot111, F = 100 \text{ cm}^2$$

$$v_- = 3\cdot07 \times 10^5 \ E/p; \ v_+ = 8\cdot78 \times 10^3 \ E/p.$$

The discharge is started at the time $t = 0$ by a pulse of light giving 10^2 initial electrons at the cathode: $j_F(t) = q \cdot \delta(t), q = 10^2 \ q_e$. This example has already been used for the simple theory.[4]

2. Results of the calculation

Using the method of calculation described in Appendix 1, the field strength, the current densities j_+ and j_- and the charge densities n_+ and n_- were calculated for a series of moments t_ν. The following stages of discharge could be distinguished:

A. Formation of initial space discharge $\nu = 1 \ldots 20$
B. Development of the cathode fall $\nu = 21 \ldots 30$
C. Rise in current due to the ionizing space wave $\nu = 31 \ldots 50$
D. Plasma with limited conductivity $\nu = 51 \ldots 62$

A survey of this time interval will be seen in *Figure 1*; this shows the current components $J_+(t)$ and $J_-(t)$ as well as the total current $J_+(t) + J_-(t)$. We shall now go on to discuss the individual stages in detail:

Formation of initial space discharge (A)

The current densities of the electrons j_- and the ions j_+ are about the same size here (*Figure 2*). The predominant space charge is that of the positive ions n_+; the ions remain longer within the discharge gap because of their lower drift velocity. Because of the initial structure of the positive space charge, with a maximum in front of the anode (*Figure 3*) the field strength decreases at the anode and builds up at the cathode. The structure of the positive space charge is maintained over a large time interval although its size is increased by about two powers of ten.[2] This can be explained by the steep rise in the current (as a result of this the ion migration can be ignored), and in particular by the change in the electrical field, within which the zone of production of the ions is displaced in the direction of the cathode: in this way the ion distribution typical of the case without space discharge is produced by the interaction between production and migration.

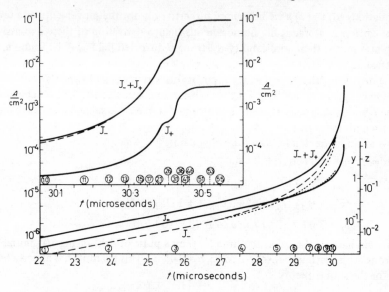

Figure 1. Variation with time of the current components J_- and J_+ and the total current $(J_- + J_+)$. The times t_ν are marked by circles. The dotted line in the lower part of the diagram is the course as calculated from reference 4; the scales on the right-hand side give the standardizations $y = K_+J_+$ and $z = K_-J_-$. The current curve beginning at $t = 0$ is shown in *Figure 5* of reference 4.

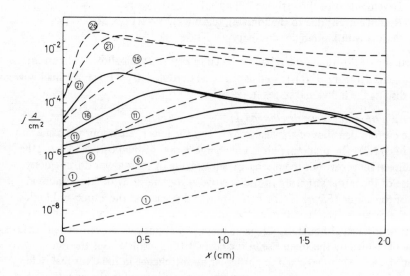

Figure 2. Current densities $j_+(x)$ (———) and $j_-(x)$ (– – – –) for various times t_ν; the values of ν are marked by circles here and in the later diagrams.

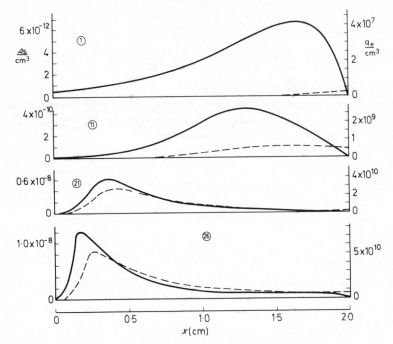

Figure 3. Space charge densities $n_+(x)$ (———) and $n_-(x)$
(– – – –) for various times t_ν.

The development with time undergoes considerable acceleration during the A stage; the time constant rises by a factor of about 20 as compared with the rise in the absence of a space charge (which is determined by $\mu_{ph} = 1\cdot111$) (*Figure 1*).

Development of the cathode fall (B)

Towards the end of the A stage, a region of high field strength is concentrated ahead of the cathode (*Figure 4*). As a large electron current is flowing in this region, the ionization production αj_- (i.e. the number of ions or electrons produced per unit time and volume) is correspondingly high. There is nearly no production of ions and electrons in the rest of the region.

In particular, the majority of the photons that release secondary electrons at the cathode are produced in the zone in front of the cathode. We can speak of a cathode fall, in line with the processes that occur during glow discharge.*

During stage B the cathode fall is concentrated into a narrow zone (*Figure 4*). At the same time the electron current leaving the cathode fall becomes noticeable.

---

* There are the following differences between this and the glow discharge process: (a) the secondary mechanism is the γ_{ph} effect, ions that can give the γ_i effect reach the cathode at a later point in time. (b) the current density in this case is well below the current density of a normal cathode fall. Thus the cathode fall described here is inefficient and entails a much higher voltage drop than does a normal glow discharge.

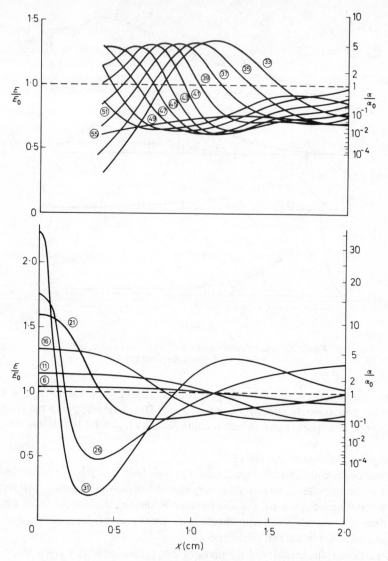

Figure 4. Course of the field strength $E(x)$ and the coefficient
$\alpha(x)$ for various times t_ν.

An excess negative space charge builds up particularly on the anode-side boundary
of the cathode fall. The associated decrease in field strength limits the outflow of
electrons from the cathode fall.

A detailed calculation of the cathode-fall region would be very involved, since
a finer x division than in the rest of the region is needed. In our further studies
$(\nu \gtrsim 34)$ we shall exclude the cathode-fall region $(x \leqslant x_K = \frac{4}{25}d)$; we need then

only consider the cathode fall from the point of view of an adequate supply of electrons. We take

$$\int_0^{x_K} E\,dx = 0.15\,U_0$$

as the voltage drop needed for the cathode fall.

Ionization wave (C)

After the cathode fall has been formed, an intensive current of electrons flows into the rest of the discharge gap and causes a change in the field relationships in the discharge gap. A significant ionization production αj_- appears first at the anode, partly because the field strength is highest here and partly because the electron current has a maximum here.

It is interesting to find that this ionization process continues as a wave in the direction of the cathode (*Figure 5*). The velocity of this wave, 2.8×10^7 cm/s, is five times higher than the drift velocity of the electrons. After the wave has passed, both the electron and ion density have increased by a factor of four.

The ionization wave represents a mechanism that brings about a rapid increase in current in the column of plasma. An essential condition for the appearance of the ionization wave is the inflow of a sufficient number of electrons to the cathode-side front of the wave; under our scheme these electrons are produced in the cathode fall. The mechanism of the ionization wave, particularly the simultaneous

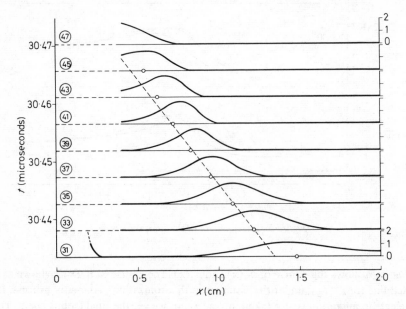

Figure 5. Ionization waves, αj_- is shown (in A/cm³) for various times t_ν. The circles on the abscissae mark the centre of gravity of the curve concerned. Dotted line: $v = 2.8 \times 10^7$ cm/s.

appearance of the field strength maximum, can be appreciated from *Figure 6*, which shows the individual functions for two consecutive moments t_{37} and t_{39}.

Two processes are responsible for the variation of $n_-(x)$ and $n_+(x)$ with time. The first is the production of ions and electrons by collision ionization and the second is the motion of electrons (the movement of ions can be ignored in this case). A change in the resulting space charge is due exclusively to the motion of the electrons, since electrons and ions are produced in equal quantities. This can be seen at once if we use the Townsend differential equations (1) and (2) to give

$$\frac{\partial(n_- - n_+)}{\partial t} = -\frac{\partial j_-}{\partial x}$$

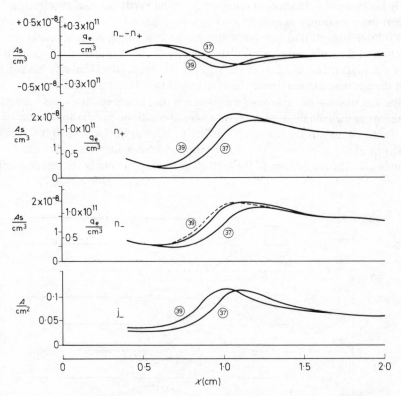

Figure 6. Development of the ionization wave.

Figure 6 shows the course of $n_-(x)$ and $n_+(x)$. The course of $n_-(x)$ is shown as a dotted line for $t = t_{39}$, under the assumption that ionization is the only process. If the electron migration is also taken into account, we get the final (solid) curve. The resulting space charge $n_-(x) - n_+(x)$ is shown in the upper part of *Figure 6*; the displacement in the direction of the cathode is plainly visible here; according to the Poisson equation this will then bring about a simultaneous migration of the field strength maximum.*

The passage of the ionization wave is accompanied by a rise in the total current (mainly an electron current) in the discharge gap. Because the discharge gap is linked to the external circuit, this brings about a fall in voltage (*Figure 7*). As a rule, when $\nu > 45$, a second ionization wave starting from the anode is produced.

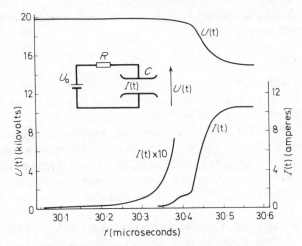

Figure 7. Variation with time of the current $I(t)$ and the voltage $U(t)$ in the discharge gap. Values for the equivalent-circuit diagram: $R = 485 \ \Omega$, $C = 28$ pF.

The amplitude of this second wave is much smaller than that of the first one, since the voltage of the discharge gap has fallen off considerably in between. This second wave is of no importance for what follows.

Plasma with limited conductivity (D)

Once the ionization wave has reached the cathode fall, a plasma has been produced in the discharge gap between x_K and d which has no special structure as regards the ion and electron densities. The field strength is approximately constant with x for $\nu = 55$ (*Figure 4*). At the same time the rise in current (or the decrease in voltage $U(t)$) comes to a standstill; the field strength has reached a value at which only slight ionization can take place.

The processes in the approximately stationary discharge are determined by the size of the positive space discharge n_+. The changes in field strength as given by the Poisson equation cause the negative space charge (in the form of rapidly moving electrons) to be the same as the positive space charge: $n_- \approx n_+$. The current

---

* Probably the ionization wave described here involves the same process described elsewhere (Westberg, R. G., *Phys. Rev.* **114**, 1, (1959)) as a current-increasing mechanism ('return arc plasma').

transport is carried out chiefly by the electrons: $j_- = v_- n_-$, whereas the ion component j_+ of the current is smaller, in proportion to the mobilities. The plasma that fills the space between the cathode fall and the anode has a limited conductivity, because n_+ is invariable.

As the discharge proceeds beyond $\nu > 55$, two processes are of importance: (1) the production of an anode fall, (2) residual ionization, through which a slow rise in n_+ (and a corresponding rise in conductivity) is produced. The anode fall is completely developed at about $\nu = 57$ ($t = 31$ μs); it then takes over the progressive regeneration of the migrating ions.

3. Discussion

Comparison with experiment

We can check the result of our theory against a measured course of the voltage $U(t)$. For this purpose we have available an oscillogram taken at $\Delta U/U_D = 2.5\%$. The difference in overvoltage is not important, as explained in the introduction. The comparison shown in *Figure 8* reveals that the calculation reproduces the measurements: in the region $t > 30.5$ μs (stage D) there is an approximately constant difference between measurement and calculation. This difference can probably be ascribed to the fact that the cathode fall needs a higher voltage drop than 0.15 U_0.

The present calculation shows that the further development of the Townsend discharge under the conditions given here is bound to lead to a partial breakdown of the voltage. An earlier explanation[11] located the partial breakdown in the spark channel stage ('transverse instability'); an estimate of the energy involved in the discharge reveals however that it cannot be limited to a narrow channel in this case.[6] Under certain conditions partial breakdown can also occur in a channel

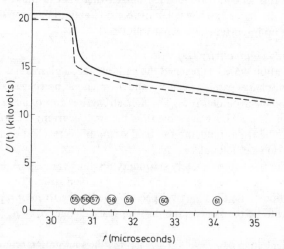

Figure 8. Course of voltage in the discharge gap, —— measured, – – – calculated. The time axis relates to the calculated curve.

structure, i.e. when the energy supplied to the discharge gap C is not sufficient for the formation of a high-conductivity spark channel. The energy flowing through R will then give rise to total breakdown. This behaviour can be observed at low pressures in particular.[12]

Transition to normal glow discharges and electric arcs

By analogy with the stationary discharge processes we should also expect the Townsend discharge to be changed into a glow discharge during ignition. The discharge in stage D, with its cathode fall, column of plasma and anode fall, has the structure of a glow discharge, but differs from a normal one by its lower current density and its higher voltage requirements. The data for a normal glow discharge in hydrogen at $pd = 500.2$ torr cm are:[13] current density 16 A/cm^2, ignition voltage 2.5 kV.

As the experiments have shown, the discharge after stage D can become an electric arc (spark). This process corresponds to the glow-arc transition in stationary discharges.

The present results relate to hydrogen and nitrogen. In air, a mechanism that is not yet quite clear causes the discharge to alter after a relatively short time from stage D to an electric arc. This gives the typical stepwise oscillogram.[5, 6] At high pressures the time that elapses before voltage breakdown is so short that it is often very hard to distinguish the stages.

Appendix 1. The numerical calculation

For the general case, where the field strength E and the associated coefficients α, v_- and v_+ are arbitrary functions of position x and time t, the following expressions are valid:

$$j_-(x, t) = j_-(\xi, \theta_-(\xi)) \exp [S(\xi)] \left\{ \frac{\partial \theta_-(\xi)}{\partial t} \right\} \tag{1.1}$$

$$j_+(x, t) = j_+(\xi, \theta_+(\xi)) \left\{ \frac{\partial \theta_+(\xi)}{\partial t} \right\}$$
$$+ \int_x^\xi \alpha(\xi', \theta_+(\xi'))j_-(\xi', \theta_+(\xi')) \left\{ \frac{\partial \theta_+(\xi')}{\partial t} \right\} d\xi' \tag{1.2}$$

The auxiliary functions $\theta = \theta_-(\xi)$ and $\theta = \theta_+(\xi)$ (or the inverse functions $\xi = \xi_-(\theta)$ and $\xi = \xi_+(\theta)$) describe the path of an electron or ion that reaches the position x at the time t:

$$\frac{d\xi_\pm}{d\theta} = \mp v_\pm(\xi_\pm, \theta) \quad \text{with} \quad \xi_\pm |_{\theta=t} = x \tag{1.3}$$

A further auxiliary function $S(\xi)$ (gas amplification along the electron track from ξ to x) is defined by

$$S(\xi) = \int_\xi^x \alpha(\xi', \theta_-(\xi'))d\xi' \tag{1.4}$$

With the help of the relationships:

$$\frac{\partial \theta_\pm}{\partial t} = \pm v_\pm(x, t)\, \frac{\partial \theta_\pm}{\partial x}, \tag{1.5}$$

$$\frac{\partial S}{\partial t} = v_-(x, t)\, \left\{ \alpha(x, t) - \frac{\partial S}{\partial x} \right\} \tag{1.6}$$

it can be shown that equations (1.1) and (1.2) satisfy the Townsend equations, numbers (1) and (2).

Apart from the factors $\partial \theta_-/\partial t$ and $\partial \theta_+/\partial t$, equations (1.1) and (1.2) correspond to the expressions for a static discharge,[14] or those derived earlier for the special case of constant drift velocity.[3,4] The additional factors $\partial \theta_-/\partial t$ etc. are needed to describe the variations in current density brought about by the periodic changes in the transit time of the electrons or ions.

References

 1. Bandel, H. W., *Phys. Rev.* **95**, 117, (1954); Kluckow, R., *Z Phys.* **148**, 564, (1957); Mielke, H., *Z. Angew. Phys.* **11**, 409, (1959); Kluckow, R., *Z. Phys.* **161**, 353, (1961); Schlumbohm, H., Dissertation, Hamburg, (1961); Hoger, H., *Dielectrics* **1**, 94, (1963).
 2. Pfaue, J., *Z. Angew. Phys.* **16**, 15, (1963).
 3. Köhrmann, W., *Z. Angew. Phys.* **11**, 414, (1959).
 4. Köhrmann, W., *Z. Naturforschg.* **19a**, 245, (1964).
 5. Köhrmann, W., *Z. Angew. Phys.* **7**, 187, (1955); *Ann. Phys.* **18**, 379, (1956); Mori, T., *Electrotechn. J. Japan* **2**, 54, (1956); Miyoshi, Y., *Bull. Nagoya Inst. Technol.* No. 8, (1956).
 6. Dehne, K., Köhrmann, W., and Lenné, H., *Dielectrics* **1**, 129, (1963).
 7. Saxe, R. F., *Brit. J. Appl. Phys.* **7**, 336, (1956); Tholl, H., *Z. Naturforschg.* **19a**, 346, (1964).
 8. Ward, A. L., VIth Int. Conf. Ioniz. Phen. in Gases, Paris, (1963).
 9. Börsch-Supan, W. and Oser, H., *J. Res. Nat. Bur. Stand.* **67B**, 41, (1963).
10. Kapzow, N. A. Elektrische Vorgänge in Gasen and im Vakuum, 1955 (Deutscher Verlag der Wissenschaften, Berlin).
11. Rogowski, W., *Z. Phys.* **100**, 1, (1936).
12. Gänger, B., *Arch. Elektrotechn*, **39**, 508, (1949); Der elektrische Durchschlag von Gasen, Springer-Verlag, Berlin, (1953).
13. v. Engel, A., and Steenbeck, M., *Elektrische Gasent Ladurgen*, Vol. 2. Springer-Verlag, Berlin, (1934).
14. Schumann, W. O., *Z. Techn. Phys.* **11**, 131, 194, (1930); Crowe, R. W., Bragg, J. K., and Thomas, V. G., *Phys. Rev.* **96**, 10, (1954); Ward, A. L., *Phys. Rev.* **112**, 1852, (1958).

Introduction to Paper 16 (Braginskii, 'Theory of the development of a spark channel')

If space permitted, it would have been desirable to give, in addition to this paper of Braginskii's, the earlier paper by S. I. Drabkina (*J. Exptl. Theoret. Phys.* **21**, 473, (1951)) on the radial expansion of spark channels. This is particularly so since Braginskii's paper is in some ways an extension of Drabkina's, the main difference in the two treatments of the problem being that Drabkina gives the radius of the channel and the velocity of its growth as a function of the energy released in the channel, whereas Braginskii considers a particular mechanism for the process and gives the channel radius as a function of the current flowing.

As described in paper 14, various workers have found Braginskii's theory to give better agreement with experiment than Drabkina's, which appears to over-estimate the radii of the spark channel and associated shock wave by about a factor of two. Several objections to her basic assumptions have been made to account for the discrepancies. However, a recent paper by Hallgren and Sigmond at the Ninth International Conference on Phenomena in Ionized Gases, Bucharest (1969), pointed out that there is an error in the constants tabulated by Drabkina for hydrogen. When this error was corrected Hallgren and Sigmond found good agreement between Drabkina's theory and their experimental results and those of Koppitz. Further, there was little difference between the predictions of the corrected Drabkina theory and that of Braginskii. The present agreement between experiment and theory therefore appears to be very satisfactory.

Further Reading

Drabkina, S. I., *J. Exptl. Theor. Phys.* **21**, 473, (1951).
Hallgren and Sigmond, Ninth International Conference on Phenomena in Ionized Gases, (contributed papers) Bucharest (1969).
Koppitz, J., *Naturforsch* **22a**, 1089, (1967).

16

Theory of the development of a spark channel

S. I. BRAGINSKII*

1. Introduction

In the present paper we consider the development of a spark channel under comparatively high pressures and moderate currents. This process has been studied in detail by Mandel'shtam and his co-workers.[1-6] In reference 1, on the basis of experimental results, the idea was expressed that the rapid development of a spark channel is accounted for by the excitation of a shock wave. In subsequent papers, this phenomenon was studied in detail, both experimentally and theoretically. The theory of the development process was given by Drabkina; the results of her calculations are in good agreement with experiment. However, the theory advanced by Drabkina is not complete; the electrical conductivity and the temperature in the channel are not computed in this theory, so that it does not permit us to calculate the parameters of the spark directly, by starting from the law of current growth. Rather, it only relates the velocity of its growth with the energy released in the channel; this latter energy must be determined experimentally.

In the present research, an attempt is made to consider a specific mechansim of the discharge and to construct a step-by-step theory of the development of the channel, with account of the electrical conductivity and the thermal conductivity of the ionized gas in the channel.

In accord with the results of references 1 to 6, the picture of the development of the spark channel can be represented in the following form. A comparatively narrow current-carrying channel is formed in the gas, with high temperature and ionization. Joule heat is released in this channel, which then leads to an increase in the pressure and a thickening of the channel. The thickened channel acts like a piston on the remaining gas and, since the expansion takes place with supersonic speed, it produces a shock in the gas; this shock is propagated in front of the original 'piston'. The temperature in the vicinity of the shock (between the wave front and the 'piston') is much higher than in the gas at rest, and the temperature

* Reproduced from *Soviet Physics, JETP*, 7, 1068, (1958). Original paper in *J. Exptl. Theor. Phys. (U.S.S.R.)* 34, 1548, (1958) by permission of The American Institute of Physics, New York.

in the channel itself is still many times higher than in the shock wave. Consequently, the density of the gas in the channel is very low, and the major part of the mass of the moving gas is displaced from it, which also makes it possible to consider the boundary of the channel as a piston.

The very fact of the formation of the narrow channel can evidently be understood by starting from the following considerations. After the gas sparks over and becomes conducting, Joule heat is released at points of flow. As is well known, the electrical conductivity of the gas increases rapidly with temperature. Thus, at a high degree of ionization, when the collisions of electrons with ions are important, the electrical conductivity is proportional to $T^{3/2}$, while at low ionization this dependence is even stronger, (because of the fact that the degree of ionization increases rapidly with temperature). As a consequence, a tendency appears toward a concentration of current in a comparatively narrow channel, so that at the places where the temperature is higher, the conductivity is also great, a large current exists there, and a large amount of heat is liberated, which leads to more heating, etc. The physical processes which determine the breadth of the channel and limit the concentration of current are the leakage of heat from the channel and the broadening of the heated region under the action of the pressure.

With some indefiniteness, we can consider as the channel the region from the axis to the point where the temperature becomes so low that ionization begins to fall off appreciably. In the channel, we can neglect the inertia of the gas, but it is necessary to take into consideration the release and transfer of heat. In the shock-wave region, the inertia must be considered, but we can neglect the electrical and thermal conductivities. These two regions as separated by a transition layer, the 'shell' of the channel. Heating and ionization of the gas that enters the channel take place in the shell.

2. Fundamental equations

The fundamental equations of the problem under consideration are the equation of continuity, the equation of motion, and the equation of energy transfer:

$$\frac{\partial \rho}{\partial t} + v \frac{\partial \rho}{\partial r} + \rho \frac{\partial (rv)}{r \partial r} = 0; \tag{2.1a}$$

$$\rho \left(\frac{\partial v}{\partial t} + v \frac{\partial v}{\partial r} \right) + \frac{\partial p}{\partial r} = 0; \tag{2.1b}$$

$$\frac{\partial}{\partial t} \left(\rho \epsilon + \frac{\rho v^2}{2} \right) + \frac{1}{r} \frac{\partial}{\partial r} \left\{ r \rho v \left(\epsilon + \frac{p}{\rho} + \frac{v^2}{2} \right) \right\} + \frac{\partial (rq)}{r \partial r} = jE. \tag{2.1c}$$

Here ρ is the density, v the velocity, p the pressure, ϵ the internal energy per unit mass of gas, q the heat flow, j the current density, and E the electric field.

We shall write the equation of state in the form.

$$p = (n_e + n_i)T = (Z + 1)\rho T/m_a, \tag{2.2}$$

where m_a is the average atomic mass, n_e and n_i the number of electrons and ions per unit volume, Z the average ionic charge, and $n_e = An_i$. The temperature is expressed in energy units.

We shall assume that the ionization in the channel can be computed by Sach's formula. This problem is considered in detail in reference 6.

The internal energy of the gas in the channel is expressed in the form

$$\epsilon = \frac{3}{2}\frac{p}{\rho} + \frac{I}{m_a} = \frac{p}{\rho}\left[\frac{3}{2} + \frac{I}{(Z+1)T}\right], \tag{2.3}$$

where I is the total energy of ionization plus the energy of dissociation, referred to a single atom. It is appropriate to apply equation (2.3) in the case of complete ionization, for example, for hydrogen, $Z = 1, I = 15\cdot74$ eV. For incomplete ionization, the energy of ionization increases with increasing temperature. According to Sach's formula, the ratio I/T depends rather weakly on the density and temperature; therefore, for a not too wide an interval of change of these parameters, the expression in the square brackets can be considered to be approximately constant. In this case it is more suitable to take the expression for the energy in the form

$$\epsilon = \frac{1}{\gamma - 1}\frac{p}{\rho}, \tag{2.3a}$$

as was done by Drabkina.[2] Here γ is the effective ratio of specific heats. The value of γ is somewhat different for the gas in the channel and in the shock wave. According to reference 2, $\gamma = 1\cdot25$ for hydrogen and $1\cdot22$ for air.

Transfer coefficients

The electrical conductivity σ and the thermal conductivity κ differ strongly for the ionized gas (see, for example, reference 7):

$$\sigma = \sigma_1(Z)\, T^{3/2} = 3\sigma'\,(Z)\, T^{3/2}/4e^2\sqrt{2\pi m\lambda}; \tag{2.4}$$

$$\kappa = \kappa_1(Z)T^{3/2}. \tag{2.5}$$

Here e and m are the charge and mass of the electron, $\lambda = \ln(3T^{3/2}/Ze^3\sqrt{4\pi n_e})$, and $\sigma'(Z)$ is a dimensionless coefficient. For $Z = 1, 2, 3$ and 4 we have, respectively, $\sigma' = 1\cdot95, 1\cdot135, 0\cdot840$, and $0\cdot667$. The value of $\kappa e^2/\sigma T$, according to the Wiedemann-Franz law, is of the order of unity. For $Z = 1, 2, 3$ and 4 this combination is equal to $1\cdot62, 2\cdot16, 2\cdot40$ and $2\cdot60$ respectively. The 'Coulomb logarithm' λ is only slightly sensitive to the values of the quantities entering into it. For $\lambda = 5$, for example, we have $\sigma_1(1) = 3\cdot4 \times 10^{-13}\ \text{s}^{-1}\ \text{eV}^{-3/2}$, and $\kappa_1(1) = 3\cdot9 \times 10^{20}\ \text{cm}^{-1}\,\text{s}^{-1}\,\text{eV}^{-5/2}$.

We note that the electrical conductivity of air increases with the temperature more slowly than $T^{3/2}$, because of the increase of Z as a consequence of ionization. At a temperature on the order of several electron volts, it changes approximately as $T^{1/2}$ and is equal to $2 \times 10^{14}\ \text{s}^{-1}$ for $T \approx 3$ to 4 eV.

Radiation

A simple estimate, taking experimental data[3 -5] into account, shows that if the radiation from the channel were black-body radiation, it would carry several ten-fold more energy than is actually released in the channel. In fact, the radiation is nonequilibrium and flows freely from the channel. For open radiation, div. $q = Q_R'$, where Q_R' is the energy radiated per unit volume. The fundamental mechanism of open radiation is the retardation radiation

$$Q_{ret}' = 1\cdot5 \times 10^{-25}\, n_i n_e Z^2 T_{ev}^{1/2}\ (\text{erg-cm}^{-3}\,\text{s}^{-1}) \qquad (2.6)$$

(see reference 8) and recombination radiation. For hydrogenlike atoms, the latter can be computed from the approximate formula

$$Q_{rec}' = 5 \times 10^{-24}\, n_i n_e Z^4 T_{ev}^{-1/2}\ (\text{erg-cm}^{-3}\,\text{s}^{-1}). \qquad (2.7)$$

Equation (2.7) was obtained by V. I. Kogan, using the cross section of recombination at the different levels given in reference 9.

For partially ionized atoms of the different elements, which cannot be considered hydrogenlike, one must expect that the radiation of Z charged ions is greater than calculated by the 'hydrogenlike' formulas, because of the incomplete screening. This circumstance can be taken into consideration if we use the effective charge $Z_{eff} = Z + \Delta$ in place of the actual charge of the ion Z. According to Unsöld,[10] we can take as a sort of mean value (as a rough approximation, with great uncertainty) $\Delta = 1\cdot5$.

Radiation in the discrete spectrum as a result of resonance absorption is forbidden for many lines and is close to the equilibrium (Planckian) in intensity. The presence of such radiation increases the thermal conductivity of the plasma. In the present research, however, we shall not consider this radiation thermal conductivity, since its calculation is a complicated independent problem which requires a detailed knowledge of the spectrum and the line widths.

Skin effect and the magnetic field

The penetration depth δ of the field after a time t can be estimated by the formula $\delta^2 \sim c^2 t/2\pi\sigma$. According to Abramson and Gegechkori,[3, 4] $\sigma = 2 \times 10^{14}\,\text{s}^{-1}$. At the instant $t = 10^{-6}\,\text{s}$, the radius of the channel becomes $a \sim 1$ mm, which yields $\delta^2/a^2 \sim 10^2$. Thus we can consider the electric field to be constant over the cross section and use the expression $j = \sigma E$ for the current density.

We now estimate the role of magnetic forces, for which we compare the magnetic pressure $H^2/8\pi$ with the gas-kinetic pressure. The latter has the same order of magnitude as the kinetic energy per unit volume. If an appreciable amount of the liberated Joule heat remains in the form of the kinetic energy of particles in the channel, then the pressure can be estimated as

$$p \sim \frac{1}{\pi a^2}\frac{J^2 t}{\pi a^2 \sigma} \sim \frac{H^2}{8\pi}\frac{\delta^2}{a^2}. \qquad (2.8)$$

It is then evident that we can regard the magnetic forces as inconsequential when we can neglect the skin effect.

The magnetic field begins to have a strong effect on the kinetics of the electrons (on the electrical conductivity and especially on the thermal conductivity) when the frequency of their rotation in the magnetic field $\omega = eH/mc$ is comparable with the collision frequency $1/\tau$. For typical values of the magnetic field and density in the channel, Mandel'shtam and his co-workers obtained in their experiments $\omega\tau \sim 0 \cdot 1$.

We shall neglect both the magnetic forces and the effect of the magnetic field on the kinetics of the electrons.

The shell of the channel. Ionization jump

Since we are dealing with entirely different simplifications in the channel region and the shock-wave region, it is necessary to establish the condition of joining the solutions on the boundary between the two regions. Physically, the joining takes place in the transition region, in the shell of the channel, where a transition occurs from a strongly ionized gas to a weakly ionized one, and where an intense ionization process is taking place. We shall not investigate the behaviour of the quantities in the transition layer, but shall consider them (approximately) as discontinuities, as is usually done for discontinuities in hydrodynamics. We shall assume here that the transition region is not very wide.

We shall denote by $\dot{a}$ the velocity of motion of the discontinuity, and use the index 1 for quantities on the outside and the index 2 for quantities on the inside (the channel side). The laws of conservation of mass and momentum take the form

$$\rho_1(v_1 - \dot{a}) = \rho_2(v_2 - \dot{a}) \equiv g; \tag{2.9a}$$

$$p_1 + \rho_1(v_1 - \dot{a})^2 = p_2 + \rho_2(v_2 - \dot{a})^2. \tag{2.9b}$$

The density in the channel is very low, $\rho_2 \ll \rho_1$; therefore, the first condition yields $v_1 = \dot{a}$. The pressure jump is expressed in the form $\Delta p = p_1 - p_2 = g^2 (\rho_2^{-1} - \rho_1^{-1})$. Considering $\rho_2 \ll \rho_1, g \sim \rho_2\dot{a}, p \sim \rho_1 a^2$, we find that the pressure jump is small:

$$\Delta p/p \sim \rho_2/\rho_1 \ll 1. \tag{2.10}$$

We shall assume that the pressure does not undergo a jump. Then, neglecting the heat flow on the side of the cold dense gas, we obtain the condition

$$g(\epsilon_2 + p/\rho_2) + q_2 = 0. \tag{2.11}$$

from the conservation of energy.

3. Quasi-self-similar solution*

As is well known (see reference 11), the motion described by two-dimensional parameters is self-similar. In our case, the motion as a whole depends on a large

---

* Referred to elsewhere as the 'auto-modelling' solution.

number of parameters; however, we can find an approximate solution which is self-similar separately in the region of discharge and in the region of the shock wave. Curves representing the dependence of different quantities on the radius remain, in each of the regions, the same with passage of time, but their scales change in each region according to its own law.

Shock wave

The motion of the gas outside the channel is completely determined if the time dependence of the radius of the channel is given. The boundary of this channel plays the role of a piston which displaces the gas. If this dependence has a simple power form

$$a(t) = At^k \tag{3.1}$$

and if the pressure in the wave is so large that the pressure of the undisturbed gas can be neglected, then the motion in the region of the shock is determined by the two-dimensional parameters A, ρ_0 and is self-similar.

In place of the variables t, r in equation (2.1), we introduce the variables t, $x = r/a_c(t)$, where a_c is the radius of the wave front. We introduce also the new dependent variables:

$$\rho'(x) = \rho/\rho_0, \quad v'(x) = v/\dot{a}_c,$$
$$p'(x) = p/\rho_0\dot{a}_c^2 \tag{3.2}$$

Neglecting the heat released and transferred, we can rewrite equation (2.1) in the form

$$(v' - x)\frac{d\rho'}{dx} + \rho'\frac{dxv'}{xdx} = 0;$$

$$\left(1 - \frac{1}{k}\right)v' + (v' - x)\frac{dv'}{dx} + \frac{1}{\rho'}\frac{dp'}{dx} = 0; \tag{3.3}$$

$$2\left(1 - \frac{1}{k}\right)p' + (v' - x)\frac{dp'}{dx} + \gamma p'\frac{dxv'}{xdx} = 0.$$

The boundary conditions in a strong shock wave for $x = 1$ have the form

$$\rho' = (\gamma + 1)(\gamma - 1)^{-1}, \quad v' = 2(\gamma - 1)^{-1},$$
$$p' = 2(\gamma - 1)^{-1}. \tag{3.4}$$

Equations (3.3) with boundary conditions (3.4) were integrated numerically on an electronic computing machine for the values $k = 1, \frac{3}{4}, \frac{3}{5}$ and $\gamma = \frac{5}{3}, \frac{7}{5}, \frac{9}{7}$. The results of the integration are shown in *Figure 1*. The position of the piston is determined by the point where $v' = x$. The pressure at the piston p_k can be obtained from the velocity of the piston

$$p_k = K_p\rho_0\dot{a}^2, \tag{3.5}$$

where the coefficient of resistance K_p will be considered as approximately equal to 0·9 (see *Figure 1*, $K_p = p'(s)a^2/a_c^2$).

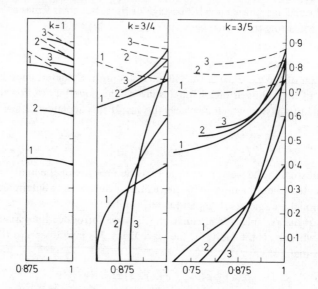

Figure 1. Distribution of the velocity v' (dashed lines), pressure p' and density ρ' behind the front of the shock wave as a function of $x = r/a_c$ for various values of k. The curves 1 correspond to $\gamma = 5/3$, 2 to $\gamma = 7/5$, and 3 to $\gamma = 9/7$.

The channel

Let us consider the case in which we can neglect radiation. Then

$$q = -\kappa dT/dr. \tag{3.6}$$

We shall assume that the temperature in the channel is appreciably higher than is necessary for complete ionization. Then, on the boundary of the channel, where the ionization is beginning to fall off, the temperature will be much less than at the centre, and we can set (approximately) $T = 0$ for $r = a$.

We transform from the variables t, r to the variables $t, s = r^2/a^2(t)$, and also introduce the new dependent variables

$$\vartheta(s) = \frac{T}{T_0}, \quad u = \frac{1}{\vartheta}\frac{r}{2a}\left(\frac{r}{a} - \frac{v}{a}\right),$$

$$y = \frac{r}{2a}\left\{\frac{q}{pa} + \frac{5}{2}\left(\frac{v}{a} - \frac{r}{a}\right)\right\}. \tag{3.7}$$

Here T_0 is the temperature on the axis. We shall regard the pressure as constant over the cross section of the channel. Equations (2.1a), (2.1c), (3.6) take the form

$$\frac{du}{ds} = \frac{1 - (1/2k)}{\vartheta}, \quad \frac{dy}{ds} = \frac{\alpha\beta}{4} \vartheta^{3/2} - \left(2 - \frac{3}{4k}\right),$$

$$\frac{d\vartheta}{ds} = -\frac{y + \frac{5}{2}u}{\alpha s \vartheta^{5/2}}, \tag{3.8}$$

where

$$\alpha = \kappa(T_0)T_0/pa\dot{a}, \tag{3.9}$$

$$\beta = \sigma(T_0)E^2 a^2/\kappa(T_0)T_0. \tag{3.10}$$

The condition (2.11) becomes $y(1)/u(1) = I/(Z+1)T_0$. It is then evident that for self-similarity the temperature ought not to depend on the time. The same applies to the quantities α, β. Making use of (3.5), (3.10), we obtain the result that $a \sim t^{3/4}$, $E \sim t^{-3/4}$. The electric field is connected with the current by the relation

$$J = \langle\sigma\rangle\pi a^2 E, \tag{3.11}$$

where the angular brackets denote an averaging over the cross section of the channel. Thus, the self-similar mode is obtained only in the case of a definite law of current increase: $J \sim t^{3/4}$. Actually, the current changes sinusoidally, but for the first quarter period of the sine wave we can use approximately the results obtained from the self-similar mode.

Equations (3.8) have been integrated numerically for $k = \frac{3}{4}$ with the boundary conditions

$$\text{for } s = 0 \quad u = 0, \quad y = 0, \quad \vartheta = 1,$$

$$\text{for } s = 1 \quad y/u = I/(Z+1)T_0, \quad \vartheta = 0(\vartheta \ll 1).$$

In the actual integration, the parameter α was assigned and β was so chosen that the quantity $\vartheta(1)$ was as small as possible. The exact value of $\vartheta(1)$ is rather unimportant, since the curves hardly depend on it over the major part of the interval. The characteristic curves are shown in *Figure 2*. The values of

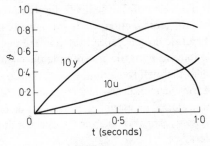

Figure 2.

$I/(Z+1)T_0 = 5·95, 2·0, 1·56, 0·9$ and $0·25$ correspond to $\alpha = 16, 8, 6·90, 5·33$ and 4 and $\beta = 1·48, 1·56, 1·60, 1·68$ and $1·80$. The coefficients $K_\sigma = \langle T_0^{3/2}/T^{3/2}\rangle$

and $K_\rho = <T_0/T>$ change in this case from 0·661 to 0·632 and from 1·49 to 1·69, respectively. We shall henceforth take approximately $\beta = 1·6$, $K_\sigma = 0·655$, and $K_\rho = 1·55$. Knowing the dependence of α and β on the temperature, it is possible to find all the parameters of the channel. However, it is more suitable to use equation (4.4), in which we must replace σ by $<\sigma> = \sigma_1 <T^{3/2}>$. In accord with (2.3), it is necessary to replace in equation (4.4′) $(\gamma - 1)^{-1}$ by $\frac{3}{2} + K_\rho I/(Z + 1)T_0$. For the temperatures of interest, the single-term approximation $\xi \approx 3\sqrt{I/(1 + Z)T_0}$ is valid with sufficient accuracy. Using equations (3.10), (3.11), (4.4) and the values of the coefficients that have been obtained, we get the radius of the channel, the temperature, and the electric field. For hydrogen, we have:

$$a = 1·53\rho_0^{-5/28}(Jt^{-3/4})^{2/7} t^{3/4};\tag{3.12}$$

$$T_k = 3·5\rho_0^{1/14}(Jt^{-3/4})^{2/7};\tag{3.13}$$

$$E = 50\rho_0^{1/4} t^{-3/4}.\tag{3.14}$$

For the temperature in the channel, the condition $T_k = (<T^{3/2}>)^{2/3}$ is assumed. Here we have expressed T_k in eV, a in millimetres, E in volts per centimetre, J in kiloamperes, t in μs, while the density unit is $0·9 \times 10^{-4}$ g/cm³. In reference 4, for a 15-kV discharge, a 2 μH self inductance and a 0·25 μF capacitance, the measured values for the radius of the channel (for hydrogen at atmospheric pressure, $\rho_0 = 1$) were 1·00, 1·55 and 2·60 mm for 0·3, 0·5 and 1·0 μs, respectively. The corresponding values computed according to (3.12) are 1·00, 1·50, and 2·45. The agreement is rather good. Experimental data for the quantities in (3.13) and (3.14) are unfortunately lacking. The existence of such data would have made it possible to verify equation (2.5) for the thermal conductivity of the plasma, since, the radiation does not play an important role in the given case.

4. Homogeneous model of a channel with a dense shell

It the removal of heat from the channel is brought about by transparent radiation, while the thermal conductivity can be neglected, then we can demonstrate a simple self-similar solution for the region of the channel: the pressure, temperature and density are constant over the cross section, while the velocity is proportional to the radius. The entire temperature drop is concentrated in the shell. The radiation is absorbed there, and the ionization of the gas entering the channel takes place in that region. If we consider the shell to be thin, we can obtain a set of equations for the basic parameters of the channel. In the general case, use can be made of these equations as a mathematical model describing, however roughly, the basic processes in the channel. In this case we can also take the thermal conductivity into consideration (approximately).

The equations for the energy balance in the channel and the shell have the form

$$\frac{dW}{dt} + p\frac{d\pi a^2}{dt} = Q_J;\tag{4.1}$$

$$\left(\epsilon + \frac{p}{\rho}\right)\frac{dM}{dt} = Q_T + Q_R. \tag{4.2}$$

where M and W are the mass and energy of the gas in the channel. Equation (4.1) is obtained by integration of (2.1c) over the cross section of the channel (including the shell) without any assumption of the form of the distribution of the quantities over the cross section. For the homogeneous model, we set $W = M\epsilon$, $M = \pi a^2 \rho$. Equation (4.2) is obtained from (2.11). The expressions for the release of Joule heat Q_J and for the heat loss by radiation Q_R and by thermal conduction Q_T can be written in the form

$$Q_J = J^2/\pi a^2 \sigma, \quad Q_R = \pi a^2 Q'_R \ (p, T),$$

$$Q_T = 1 \cdot 3 \times 2\pi \kappa T. \tag{4.3}$$

In order of magnitude, $Q_T \sim \kappa(T/a)2\pi a$, while the coefficient in (4.3) is chosen in correspondence with the results of the previous section equation (3.10) wherein $T_k = (<T^{3/2}>)^{2/3}$ was assumed for the characteristic temperature in the channel. Approximately, it can be obtained from (3.5) for a weak shock, or it can be considered equal to the pressure of the undisturbed gas, when the wave becomes weak and undergoes a transition to the acoustic type.

Making use of (3.5) and (2.3), we can rewrite (4.1) in the form

$$\rho_0 2\pi^2 a^3 \dot{a}^3 \xi = J^2/\sigma, \tag{4.4}$$

$$\xi = K_p \left[1 + (\gamma - 1)^{-1} 2^{-1} \dot{a}^{-2} \frac{d^2 a^2}{dt^2}\right] \tag{4.4'}$$

$$= K_p [1 + (\gamma - 1)^{-1}(2 - k^{-1})].$$

Here $k = \dot{a}t/a$. Comparing (4.1) and (4.2), we obtain

$$Q_T + Q_R = \eta Q_J, \tag{4.5}$$

where η is a coefficient of the order of unity. If, for example, we can neglect the change in temperature with time, then

$$\eta = \gamma \left[1 + (\gamma - 1)2\dot{a}^2 \left(\frac{d^2 a^2}{dt^2}\right)^{-1}\right]^{-1}. \tag{4.5'}$$

For a weak shock, when the pressure in the channel can be considered equal to the pressure of the undisturbed gas p_0, we get from (2.3) instead of (4.4)

$$p_0 2\pi^2 a^3 \dot{a}\gamma/(\gamma - 1) = J^2/\sigma. \tag{4.6}$$

Equation (4.5) retains its form, but the coefficient η will be different. For example, if we neglect the change in temperature with time, then we have simply $\eta = 1$ in place of (4.5). Equations (4.4), (4.5), together with (4.3) and (3.5), (3.11) allow us to find all the parameters of the channel.

Let us consider the channel in air. The conductivity $\sigma(T)$ for air in the temperature range of interest to us changes comparatively slowly (see section 3) and, by (2.4), can be taken to be approximately $\sigma = 2 \times 10^{14}$ s^{-1}. This is supported by the experimental data. If, making use of references 3 and 4, we take the electrical conductivity into account, then it is shown that within wide limits of change of the parameters of the discharge, σ does not depart appreciably from this value. Assuming $K_p = 0.9$, $\gamma = 1.2$ and $J \approx t$, we get $\xi = 4.5$. For these values of σ and ξ, we get for the channel radius (from equation (4.4))

$$a = 0.93\, \rho_0^{-1/6} J^{1/3} t^{1/2}. \qquad (4.7)$$

Here a is in millimetres, J in kiloamperes, t in microseconds, and we take as the density unit the density of the air at atmospheric pressure, 1.29×10^{-3} g/cm^3.

The experimental values of the radius[4] for a discharge voltage of 15 kV and capacitance $C = 0.15$ μF, at 0.3, 0.5 and 1.0 μs, are the following: for a coil inductance of $L = 2$ μH (which corresponds to $\dot{J} = V/L = 7.5 \times 10^9$ A/s): 0.65, 0.95 and 1.55 mm, respectively; for $L = 12$ ($\dot{J} = 1.25 \times 10^9$), 0.33, 0.50 and 0.80 mm, respectively; for $L = 64$ ($\dot{J} = 2.4 \times 10^8$), 0.18, 0.25 and 0.40 respectively. The corresponding values computed from (4.7) are 0.67, 1.0 and 1.62 (for $L = 2$); 0.35, 0.57 and 0.99 (for $L = 12$), 0.21, 0.32, and 0.58 (for $L = 64$). The agreement is excellent. A certain saturation at larger self inductances and values of the time is explained by the fact that (4.4) does not take the initial pressure into account. If this were done in (4.1), for example, by means of the interpolation $p = K_p \rho_0 a^2 + p_0$, then the agreement with experiment would be improved.

The spark discharge in air has also been investigated experimentally by Norinder and Karsten.[12] The values of the radius computed by (4.7) agree satisfactorily with their experimental data.

The temperature in the channel can be calculated by (4.5) and (4.3). However, this computation is difficult in practice because of the absence of reliable data on the radiation of air. We shall only put down some estimates. The coefficient η is of the order of unity. For the same discharge which was considered previously,[3] at $t = 1$ μs, the Joule heat (for $L = 2$, 12 and 64 μH) is $Q_J = 1.7 \times 10^{13}$, 3×10^{12}, 4.2×10^{11} erg/cm s, respectively. For $L = 64$, using (4.5), we obtain $T = 3.7$ eV, while all the heat is transferred by the electronic conductivity; we can neglect radiation ($Q_R \approx 10^{10}$ erg/cm s). For $L = 12$, the thermal conductivity and radiation have the same order of magnitude, but for $L = 2$, the heat is primarily conveyed by radiation. In the second case, the radiation is much greater than in the first, because of the high density of the plasma, but also because of the large value of the cross section of the channel. Taking $T = 4$ eV, we get, making use of (2.2) and (3.5), $n_i = 3.3 \times 10^{17}$, in the first case and $n_i = 9 \times 10^{17}$ in the second. These quantities greatly exceed the experimental value of 10^{17} obtained by Dolgov and Mandel'shtam.[5] According to their experimental data, $T \approx 4$ eV and $Z \approx 2$. Substituting these values in (2.5) and (4.3) we get $Q_T = 0.6 \times 10^{12}$ erg/cm s. We estimate the radiation crudely by using (2.7) with an effective charge equal to $Z + 1.5 = 3.5$. This gives

$Q_R = 1{\cdot}6 \times 10^{12}$ for $L = 12$ and $Q_R = 4{\cdot}4 \times 10^{13}$ for $L = 2$. These results correspond in order of magnitude to the experimental values of Q_J; however, the accuracy of the estimates is not very great because of the very approximate method employed in considering the radiation of air. Therefore, the role of other possible mechanisms of heat transfer, for example, radiant thermal conductivity, is not completely clear.

In conclusion, let us consider the limits of applicability of the theory developed above.

The lower limit is determined by the fact that for appreciable ionization, the temperature in the channel ought to be larger than (approximately) 1 eV. For this, the current ought to be not too small and should increase after a rather short time. The corresponding estimate can be obtained from (3.13) for hydrogen and from (4.3) and (4.4) for air. Neglecting the weak dependence on the density, and disregarding radiation, we obtain a condition for both cases, which is very rough:

$$J \gg 10^{-2} t^{3/4}. \tag{4.8}$$

The upper limit is determined by the requirement of smallness of the magnetic pressure $H^2/8\pi = J^2/2\pi a^2 c^2$ in comparison with the gas-dynamic pressure. Using (3.5) and (4.4), we get

$$H^2/8\pi p = (J/J_0)^{2/3}, \quad J_0 = (2^{1/2} K_p^{3/2}/\xi \pi^{1/2})(c^3 \rho_0^{1/2}/\sigma), \tag{4.9}$$

where J_0 is the current at which the magnetic forces begin to be appreciable. For hydrogen, setting $\sigma = \sigma_1 T^{3/2}$, $\xi = (53/T)^{1/2}$, and using (2.4), we get

$$J_0(H_2) = 50 \rho_0^{2/5} j^{-1/5}. \tag{4.10a}$$

For air, substituting a fixed value of the conductivity $\sigma = 2 \times 10^{14}$, and $\xi = 4{\cdot}5$, we get

$$J_0 \text{ (air)} = 250 \rho_0^{1/2}. \tag{4.10b}$$

The current is expressed in kiloamperes, the time in microseconds, and the density in units of $0{\cdot}9 \times 10^{-4}$ for hydrogen and $1{\cdot}29 \times 10^{-3}$ for air, both in grams per cubic centimetre.

Both criteria are well satisfied for typical cases of lightning in the atmosphere. For example, let the current of the lightning be 30 kA and the time of current flow 200 μ s, then we get $0{\cdot}55 \ll 30 \ll 250$. The form of the lightning current is not linear, so that the coefficient in (4.7) must be changed, but if (4.7) is used for a rough estimate, then we get, in the case considered, for the radius of the lightning channel $a \approx 4$ cm.

In conclusion, I express my deep gratitude to M. A. Leontovich, V. I. Kogan, D. A. Frank-Kamenetskii and S. L. Mandel'shtam for useful discussions, and to Z. D. Dobrokhotov and G. A. Mikhailov for help in setting up the program for machine computation and for carrying out the computations.

References

1. Abramson, Gegechkori, Drabkina, and Mandel'shtam, *J. Exptl. Theoret. Phys. (U.S.S.R.)* **17**, 862, (1947).

2. Drabkina, S. I., *J. Exptl. Theoret. Phys. (U.S.S.R.)* **21**, 473, (1951).
3. Abramson, I. S. and Gegechkori, N. M., *J. Exptl. Theoret. Phys. (U.S.S.R.)* **21**, 484, (1951).
4. Gegechkori, N. M., *J. Exptl. Theoret. Phys. (U.S.S.R.)* **21**, 493, (1951).
5. Dolgov, G. G. and Mandel'shtam, S. L., *J. Exptl. Theoret. Phys. (U.S.S.R.)* **24**, 691, (1953).
6. Mandel'shtam, S. L. and Sukhodrev, N. K., *J. Exptl. Theoret. Phys. (U.S.S.R.)* **24**, 701, (1953).
7. Braginskii, S. I., *J. Exptl. Theoret. Phys. (U.S.S.R.)* **33**, 459, (1957); *Soviet Phys. JETP* **6**, 358, (1958).
8. Heitler, W., *The Quantum Theory of Radiation*, (Oxford, 1954).
9. Wessel, H., *Ann. Physik* **5**, 611, (1930).
10. Unsöld, A., *Physik der Sternatmosphären* (Berlin, 1955), Ch. VI.
11. Sedov, L. I. (*Methods of Similitude and Dimensionality in Mechanics*) (GITTL, Moscow) 1954.
12. Norinder, H. and Karsten, O., *Arkiv. Mat., Astr. Fys.* **A36**, No. 16, (1949).

Introduction to Paper 17 (Penning, 'Anomalous variations of the sparking potential as a function of $p_0 d$')

In 1889 F. Paschen published the results of an experimental study of electrical breakdown in uniform fields which showed that the breakdown potential, V_s, at which a spark occurred was a function of the product (pd), p being the gas pressure and d the distance between the anode and cathode of the discharge gap. The relationship

$$V_s = \text{function of } (pd)$$

has become known as 'Paschen's Law'. The graph of V_s against pd is, of course, different for different gases and such curves are known as 'Paschen curves'.

Most of the measurements of breakdown potentials have been concerned with high values of (pd) where V_s increases with increasing pd. However, the behaviour of gases at low values of pd, where V_s increases with decreasing pd, is of considerable interest, particularly technologically since the increase of V_s at values of pd below the so-called 'Paschen minimum, $(pd)_{min}$', is often considerable for slight changes in pd. As has been pointed out by S. Haydon (*Discharge and Plasma Physics*, University of New England, 1964) and others, when the parameter E/p increases to large values at low p the drift velocity of the electrons in a discharge becomes much greater than their random velocities and the energy gained between collisions greatly exceeds the mean energy loss at a collision. The velocity distribution of the electrons becomes a function of their spatial position in the discharge gap. The usual assumptions underlying the Townsend theory of electrical breakdown obviously need to be modified. The problem has been examined theoretically by H. Neu and K. G. Müller and others and more recently by P. C. Johnson and A. B. Parker (references given below).

The paper by F. M. Penning, from which an extract is given here, was one of the first to show that the breakdown process in uniform fields below the Paschen minimum sometimes shows interesting features which are not observed at higher values of pd. The marked inflection in the Paschen curve for helium observed by Penning has been followed by studies such as those reported by L. G. Guseva, H. C. Miller, and Parker and Johnson, for the inert gases and mercury vapour. The latter authors have been able to show that their computerised Monte Carlo calculations, based on spatially-dependent ionisation coefficients and on the reflection of electrons from the anode, are capable of explaining their experimental data for mercury vapour.

Further Reading

Paschen, F., *Ann. Physik* **37**, 69, (1889).

Guseva, L. G., in *Investigations into Electrical Discharges in Gases*, edited by B. N.
    Klyarfel'd (Macmillan, New York, 1964).

Miller, H. C., *Physica* **30**, 2059, (1964) and *Int. J. Electronics*, **22**, 31, (1967).

Müller, K. G., *Z. Phys.* **169**, 432, (1962).

Neu, H., *Z. Phys.* **152**, 294, (1958).

Johnson, P. C. and Parker, A. B., *J. Phys., D,* **4**, L7, (1971).

Overton, G. D. N., and Davies, D. E., *J. Phys., D,* **1**, 881, (1968).

17

Extract from *Anomalous variations of the sparking potential as a function of p_0d.*

F. M. PENNING*

1. Introduction

According to Paschen's law the sparking potential V_s of a gas between parallel plates at a short distance is a function of $p_0 d$ (gas-pressure at $0°$C x electrode distance). The general shape of the curves $V_s = f(p_0 d)$ is shown in *Figure 1*; at high pressure the sparking potential is high because the electrons lose much energy before they reach their ionizing velocity, at low pressure the sparking potential is high because there are few ionizing collisions; at an intermediate value of $p_0 d$, $(p_0 d)_{min}$, V_s has a minimum value. Recently we have found deviations from the type of curve shown in *Figure 1*: in mixtures of neon with a small amount of argon we found two minima instead of a single one; for helium at values of $p_0 d < (p_0 d)_{min}$ the curve has the peculiar form shown in *Figure 5*.

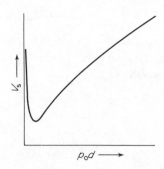

Figure 1.

* Reproduced from *Proceedings of the Royal Academy of Sciences, Amsterdam* **34**, 1305, (1931) by permission of The Royal Netherlands Academy of Sciences and Letters, Amsterdam.

2. The double minimum in the curves for neon with a small amount of argon

These measurements were performed with two types of apparatus A and B. The tube A with two iron plates of 8 cm diameter at 2 cm distance is shown in *Figure 2*;

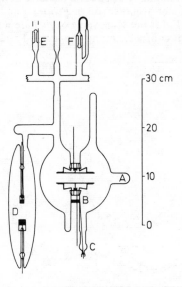

Figure 2. (Tube A).

it was intended in the first place for measurements with pure neon which will be described elsewhere.* The tube was filled with about 250 mm neon, the last small traces of impurities being removed by an arc discharge of 0·5 A with Ca-cathode in the side tube D. The desired neon-argon mixture was realized by opening one of the side tubes F which had been filled with argon beforehand. Then one of the side tubes E was sealed to the vacuum system over a liquid air trap, a manometer and a tap, the first being nearest to the tube. After shattering of the glassbulb in E the measurements could be taken.

Figure 3 shows the results of two series of measurements, one for $Ne†$ one for Ne + 0·0018% A. The curve for Ne has the normal shape, the other, however, shows two minima, which can be explained in the following way.

At large values of $p_0 d$ the sparking potential of neon is decreased very much by a small admixture of argon, because the metastable neon atoms ionize the argon atoms.[1] At low values of $p_0 d$ this effect will only have a small influence; the probability for the formation of metastable atoms is not large and those which are formed have a short life time, because of their returning into the normal state at the electrodes. Therefore in this region of $p_0 d$ the value of V_s for Ne + 0·0018% A is about the same as that for Ne and increases in about the same way with increasing

* *Physica.* To this article we refer also for further particulars as to the construction and operation of the tube.

† The neon contained about 1% helium, but the influence of this admixture on the sparking potential is small.

values of p_0d. At still larger values of p_0d the effect of the metastable atoms
becomes appreciable and V_s decreases until a second minimum is reached. This
interpretation is confirmed by other experiments with the second tube B having
also iron electrodes of 8 cm diameter but at a distance of 12 cm, mounted in a
glass tube of 10 cm diameter. Comparison of *Figures 3* and *4* shows that even with
pure neon the values of V_s for the same value of p_0d are not the same in both
cases which is in apparent contradiction to the law of Paschen. However, this law

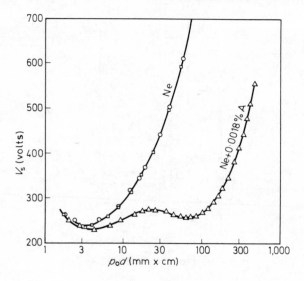

Figure 3. (Tube A, d = 2 cm).

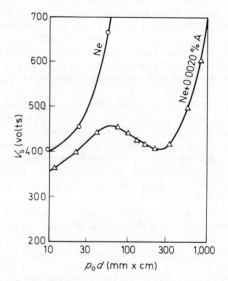

Figure 4. (Tube B, d = 10 cm).

holds only for distances of the electrodes which are small compared with their diameter, otherwise the side diffusion of electrons and ions is not negligible. In the tube B this condition is certainly not fulfilled, therefore the sparking potential is higher. Now the side diffusion of the metastable atoms is still larger than that of the charged particles which are directed by the electric field, and consequently at the same value of $p_0 d$ the fraction of metastable atoms which is destroyed at the walls is much larger in tube B than in tube A. This explains why for Ne + 0·002% A the maximum in tube B occurs at a value of $p_0 d$ which is about three times as large as that in tube A. In addition the double minimum is still more pronounced in *Figure 4* than in *Figure 3*.

3. The sparking potential of helium for values of $p_0 d < (p_0 d)_{min}$

The tube of *Figure 2* is not suited for measurements in this region because here the sparking may occur between the back sides of the electrodes. Therefore the measurements were made between the plane ends of two cylinders fitting exactly in a glass tube. Both cylinders could be moved, $p_0 d$ being varied in this case by changing d. Now for a gas pressure of 0·7 mm He the sparking potential curve for $p_0 d < (p_0 d)_{min}$ has the peculiar form shown in *Figure 5*; this curve has a minimum with respect to d.*

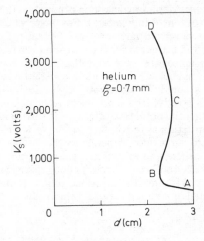

Figure 5.

In a certain range of electrode distances the sparking potential has three different values, two for increasing and one for decreasing potential difference. This effect is still more pronounced in another apparatus described in section 4.† For 2·2 cm $< d <$ 2·8 cm the current suddenly dropped to zero after reaching a certain value, the potential at the same time rising to a high value inside the region BCD (*Figure 5*), the sparking potential on CD and on BC could then be determined by increasing resp. decreasing the potential difference of the potentiometer. For $d >$ 2·8 the discharge did not stop when the current was increased; to measure in

---

* The measurements are described more fully in *Physica*, 12, 65, (1932).

† The descriptions of the apparatus and measurements referred to have not been included here. The electric field in the second apparatus was not homogeneous, but the characteristics were similar to those described here.

this region the electrode distance was at first adjusted to a small value where a high potential difference could be applied; then, with this potential difference remaining between the electrodes, the distance was increased up to the desired value.

This form of the curve for the sparking potential may be explained in the following way.

When an electron leaves the cathode it will ionize atoms on its way to the anode, the new electrons will in turn produce other ions and electrons and so on. If N is the total number of ions produced in this way by one electron leaving the cathode, then according to Holst and Oosterhuis[2] sparking will occur when $N\gamma = 1$, γ being the number of electrons liberated from the cathode per positive ion. When, for $p_0d < (p_0d)_{min}$, d is decreased, the number of atoms between the electrodes decreases and also the number of ionizations N. So, in order to fulfil the condition $N\gamma = 1$, N should be increased which can be done by increasing the potential difference. However, the probability of ionization by an electron does not increase continually with increasing electron velocity, but passes through a maximum. According to Compton and van Voorhis[3] this maximum is reached for helium at an electron velocity of about 200 volts. When, however, the potential difference between the electrodes becomes so large that the number of ionizations N is at its maximum, a decrease of p_0d cannot be compensated by increasing the potential difference: for lower values of p_0d no sparking is possible, the sparking potential curve bends to the right (point B in *Figure 5*). For larger values of d two sparking potentials occur, because a certain value of N can be realized as well to the right as the left of the maximum in the curve for the ionizing probability of the electrons.

In reality the circumstances are more complicated than was supposed in the simplified discussion given above: first γ is not a constant but increases with increasing ion velocity, secondly at sufficiently high velocity the ionization of atoms by collisions with positive ions is no longer to be neglected. These effects cause again sparking when the potential difference is sufficiently increased (part CD of the curve). The question as to whether the curve for V_s should show a minimum with respect to p_0d or not, depends on the way in which the electron building power of the positive ions increases as the ionization by the electrons diminishes. Experimentally too little is known about these elementary processes to make any predictions. With the first described tube this minimum was up to the present found only in helium although several other gases were examined.*

References

1. Penning, F. M., *Phil. Mag.* **11**, 961, (1931); *Z. Phys.* **57**, 723, (1929).
2. Holst, G. and Oosterhuis, E., *Versl. Kon. Ak. v Wet. Amsterdam* **29**, 849, (1920); **30**, 10, (1921); *Phil. Mag.* **46**, 1117, (1923); compare Penning, F. M., *Proc. Amsterdam* **31**, 14, (1928); **33**, 841, (1930).
3. Compton, K. T. and Van Voorhis, C. C., *Phys. Rev.* **27**, 724, (1926).

* *Editor's note:* In the remainder of the paper, Penning discusses the observed data in terms of the stability of the discharge.

Introduction to Papers 18 and 19 (Loeb, Extract from 'Electrical Coronas'; Loeb, 'Characteristics of the spark transition in asymmetrical field gaps for negative ion and free electron gases')

The electrical breakdown phenomena considered in papers 1 to 17 have largely been concerned with ionisation processes in uniform electric fields or in non-uniform fields produced through the action of the discharge itself. A highly interesting and complex group of discharges are those referred to as 'Corona' discharges, which may broadly be defined as discharges taking place between electrodes whose geometry produces a non-uniform field (referred to by Professor Loeb as 'asymmetrical field gaps'). The term 'corona discharge' is restricted to such discharges for which the discharge current is relatively low, i.e. it does not include the discharges obtained when complete breakdown of the discharge gap has occurred. The name 'corona' as pointed out by Professor Loeb in his book *Electrical Coronas* is derived from the French 'couronne' (a crown) which is a good description of the visual appearance of one of the many characteristic forms of the discharge.

Much of the work on corona discharges has been carried out by Professor Loeb and his colleagues and a full description of their work is given in *Electrical Coronas*. A shorter account of some of the main features of the work is given in the paper included here. As in the case of paper 13, it seems preferable to include a paper, such as this, which describes the main features of corona discharges instead of attempting the almost impossible task of selecting one or two papers from the great number published. Further, the paper is preceded by a short description of the visual appearance of the more common forms of corona although it is difficult to give in words a satisfactory picture of the observed phenomena.

Rigorous theoretical analysis of the various forms of corona discharges and of the transition from a corona discharge to complete breakdown is, of course, exceedingly difficult. The non-uniformity of the electric fields and their distortion through the accumulation of space charges complicate the analysis. Professor Nasser in his recent book *Fundamentals of Gaseous Ionization and Plasma Electronics* outlines the basic models which have been used to interpret both anode and cathode coronas under either static or impulse electric field, and chapter 11 of his book should be consulted.

One of the best known features of coronas at the cathode of a non-uniform gap subject to a static applied electric field is the occurrence of Trichel pulses (Trichel, G. W., *Phys. Rev.*, **54**, 1078, (1938)). Even in this case detailed quantitative analysis has been restricted by the lack of reliable experimental data for the very early stages of the pulses in which the development of the discharge is rapid. It is interesting to note, however, that C. Bugge and R. S. Sigmond claimed at the Ninth International Conference on Phenomena in Ionized Gases (Bucharest, 1969) that they had observed the early development of Trichel pulses to be due to a Townsend discharge.

A brief mention should be made of the wide range of applications of corona discharges. Among those listed by Loeb and Nasser are: the separation of dry ores, the discharging of static from aircraft and from surfaces used in the handling of plastics, wool and paper, and the electrical precipitation of dust. Undesirable coronas from high voltage cables and terminals are often troublesome, leading to power losses and radio and T.V. interference.

Further Reading

Loeb, L. B., *Electrical Coronas*, University of California Press, 1965.

Nasser, E., *Fundamentals of Gaseous Ionization and Plasma Electronics*, Chapter 11, Wiley, 1971.

Aleksandrov, G. N., *Soviet Phys. Tech. Phys.* **1**, 2547, (1956) and **8**, 161, (1963).

Boullard, A., *Rev. Gén. Elect.* **64**, 28, (1955).

Miyoshi, Y., *J. Inst. Elect. Eng. Japan* **78**, 141, (1958) and Chapter 4 of *Progress in Dielectrics*, Vol. 9, 1967.

18

Extract from *Electrical coronas*

L. B. LOEB*

5. Visual appearances of coronas

A few general comments might be made about the visual appearances. With highly
positively stressed electrodes, perhaps three rather characteristic forms are noted.
In the intermittent regime *just above threshold*, the most highly stressed area of an
electrode will be covered by a seemingly tightly adhering velvety glow as burst
pulses appear. This is particularly notable and becomes intense and thin at higher
pressures, especially above onset of steady corona. With coaxial cylinders it will
appear as spots at onset. As current increases with potential increase, the glow will
spread. In the intermittent regime, shortly above threshold, the glow will extend
out into the gap a millimetre or two when the pre-onset streamers appear. From
points, the glow may have a paintbrush-like appearance with a short, intensely
bright stem and a flare below. These will appear at the point tip or in spots on a
wire. The forms which at high pressures project into the gap, including the paintbrush-like manifestations, are due to the appearance of pre-onset streamers, and the
range is a rough indication of streamer advance in that gap. The tightly adhering
glow of the quiescent steady corona, beyond the pre-onset streamer potential, is a
form of a newly interpreted glow discharge with negative space charge, characteristic
of electron-attracting gases such as air, which suppresses the streamers. Generally
speaking, with or without negative ions the glow will adhere closely to the anode
near threshold. As potential increases, it will project out into the gap, especially
where the glows are restricted to patches. Reduction in pressure causes an increase
in thickness or extension of all glows into the gap roughly in inverse proportion to
the pressure.

At very high potentials approaching spark breakdown, especially in gaps in
which point radius is small compared to gap length, a narrow bundle of luminosity
composed of many superposed filamentary streamers extends well into and across
the gap along the axis. These streamers are a manifestation of incipient breakdown
streamers. They do not always appear preceding breakdown with larger electrodes
and short gaps. Reduction in pressure will broaden and diffuse all such luminosities
in three dimensions because of augmented diffusion.

For the highly stressed negative electrodes at threshold and above breakdown,
the luminosity occurs in limited spots in air and in electron-attaching gases. For a

* Reproduced from *Electrical Coronas*, University of California Press, 1965. Reprinted by
permission of The Regents of the University of California.

point of small to moderate radius, there will at onset be one spot. This discharge is an intermittent localized glow discharge of some 0·02 cm diameter near the cathode surface and extends into the gap perhaps a millimetre or two. Decreasing pressures reveal it to be complex, having a Crookes dark space, a constricted negative glow, a Faraday dark space, and a flaring positive column. This gives it a shaving brush-like appearance. The shape is dictated by the relative potential gradients along the discharge axis and in the space outside the discharge. In free electron gases at threshold, the discharges may appear first as a faint glow extending into the gap which contracts to a spot and then reverts to the initial form, oscillating between the two forms. As potentials and current increase these glows settle down at points to yield the same general configuration as for air but are more diffuse and extend farther into the gap. In some gases the spots will be cusped with the base on the electrode. As potentials increase, the spot at first increases in intensity, but presently a second spot appears, then a third. On a wire the spots multiply along the wire. In some cases they are in ceaseless motion, in others they are quiescent at one point. Both positive and negative glow patches or spots appear to, and actually do, repel each other. On larger smooth surfaces, the quiescent spots, owing to mutual repulsion and field gradients, form regular patterns. Reduction in pressure, as indicated, extends the discharges in three dimensions, roughly inversely proportional to pressure. Near breakdown in air and possibly in other gases, though not readily observed, the negative glow increases in extent, and at lower pressures the Crookes dark space and a faint general glow extending to the negative electrode are all that is seen near the point where pulses cease. Near breakdown, a very sharp luminous spike will usually appear along the axis of the spot in the fan, and with increase in potential it will extend backward toward the negative glow. At a point where this spike approaches the cathode, the spark occurs. In air in the region just referred to, the intermittent pulses characteristic of the threshold and above have ceased. When a spark-suppressing gas like freon is mixed with air, the discharge takes on a fantastic number of curious and elaborate forms which are not understood.

The spectral behaviour of breakdowns in air is of some significance because it reveals important discharge characteristics. Possibly similar inferences could be drawn from single lines or bands in other gases. Low electrical fields yield ionization and excitation that favour the arc spectrum of N_2 and nitric oxide which yields a reddish purple glow. This is characteristic of the fan in the negative point discharge and the outer areas of the paintbrush in the positive streamer region. High electrical fields produce highly excited N_2 molecules, the spectra of which are represented largely by the second positive group bands, with just a few second negative bands shown. These lie largely in the green to violet end of the spectrum and yield an intense electrical blue light. They have recently been photographed on colour film in air. This colour will be noted in the closely adhering glow of the steady positive corona, in the filamentary breakdown streamer channels, and in the axis of the streamers as well as in spark channels. the streamer tips have very high fields. The negative glow of the Trichel pulse corona in air is also quite white or bluish, as fields here are indeed high. Thus, the colour in discharge manifestations in air is a guide to field conditions. In other gases, the colours follow spectral distribution but do not clearly designate high and low field regions except possibly as they affect single lines or bands. In some gases, notably in O_2 and CO, spectra are of such low intensity in the visible as to delay the appearance of luminosity until thresholds have been exceeded by a considerable margin.

19

Characteristics of the spark transition in asymmetrical field gaps for negative ion and free electron gases

L. B. LOEB*

As a result of improvements in techniques, the last several years have witnessed a fairly complete resolution of the transition from no current, or low order conduction current, to a transient arc in both uniform field and in asymmetrical field geometry. In most cases this transition to a filamentary highly luminous arc channel is commonly called the electric spark. It has been studied in air and various other gases from atmospheric pressure down to about 50 torr. The many years of study of this problem have only recently been crowned by success because of the incredibly short time intervals and the faint luminosities of the complicated intervening steps. It required oscillographs of time resolution in the nanosecond range, reliable photomultipliers and image converter cameras, as well as intense short duration light sources for its achievement.

While the uniform field studies have been successfully carried out in the institute of Professor H. Raether[1] in Hamburg by a large succession of brilliant students, the more general sequences in the breakdown in asymmetrical gaps have been largely delineated by the workers in the writer's laboratory, of which the latest phases have just been completed.

In uniform fields at threshold for breakdown in most gases and for some gases well above it, the breakdown initiates as a Townsend type glow discharge.[2] This eventually builds up a space-charge field such that avalanches can attain a critical positive ionic space charge of the order of 10^8 ions and streamers advance towards anode and cathode respectively at higher than avalanche speeds.[3] These propagate by augmenting the applied uniform field locally by space-charge distortion and by means of photoelectric ionization of the gas in advance of the space charge. Some gases such as O_2 are readily ionized by their own photons. Air is particularly susceptible to this action since N_2 and O_2 photons ionize O_2.[4] Most other gases, unless exceptionally pure, also undergo such action in varying degrees. Cataphoretically purified argon does not appear to be susceptible to photoionization.[5]

* Reproduced from *Revue Roumaine de Physique* 13, 163, (1968) by permission of the author.

Once the streamer tips reach the electrodes, the sudden high potential gradients resulting lead to very rapidly propagating waves of potential gradient that sweep up and down the channels increasing their conductivity to the point where thermalization by ion impact can begin.[6] These waves in longer gaps may undergo successive reflections at the electrodes before the channels are sufficiently conducting.[7]

If in moist air an impulse potential 4% above the Townsend threshold, and in dry air at 6%, is applied, the spark proceeds directly without antecedent glow discharge.[8] In other gases the percent of overvoltage needed for the direct streamer spark without Townsend discharge differs and may reach 50% in argon designated spectroscopically pure. At these potentials the critical streamer forming an avalanche head of 10^8 electrons and ions is achieved at the anode, and a single anode streamer crosses the gap followed by return space wave strokes. If higher overvoltages are suddenly applied, breakdown starts with the critical streamers originating in midgap with both positive and negative streamers moving towards cathode and anode. This mechanism has been confirmed by Wagner[3] with image converter photographs. The speeds of positive and negative streamers are initially about equal and of the order of three times the avalanche speed. The speeds vary as the streamers approach the electrodes. At the time of streamer initiation the light from excited states causes liberation of new photoelectric avalanches from the cathode. When these encounter the positive streamer tip in its advance, the streamer is greatly increased in speed and luminosity. This new aspect will be again encountered in what follows.

One may now turn to the asymmetrical gaps. Only two common modifications need be discussed. One is the hemispherically capped cylindrical point-to-plane gap; the other is the coaxial cylindrical condenser. While the point-to-plane has the advantage of greater isolation of events at its highly stressed electrode, with less influence from the opposite electrode, the coaxial cylinder has the advantage of giving us an insight into the magnitude and role of space-charge effect since these are measurable and calculable. Basically, however, the general behaviour of the two types of gaps are the same. One must now distinguish essentially four situations. The first two consist of positive and negative highly stressed electrodes in gases which permit negative ions to form and in which negative-ion space charges play a role in controlling current. The second two situations employ the same electrode configurations, however in gases in which the electrons remain essentially free. These yield no negative-ion space charge in the lower or higher field regions. Air and oxygen containing or liberating gases result in negative-ion formation, as do halogen containing and certain other gases and many O_2 contaminated gases. However, in air, if pressures are sufficiently low or if temperatures exceed some $900°C$,[9] electrons remain free, as is the case in pure H_2, N_2, the inert gases, and Hg.

The breakdown sequence in air and negative-ion forming gases has been pretty thoroughly investigated and the mechanism studied down to 50 torr. In consequence of the effect of positive- and negative-ion space charges, there is usually a great potential range between corona threshold and the ultimate spark transition.[10] In both instances the negative ions are largely responsible for this action, even with a highly stressed anode. One may consider positive and negatively stressed electrodes in asymmetrical field gaps in these gases and then turn to their counterparts where electrons do not attach to yield negative ions.

1. Electron attaching gases

A. *Positive highly stressed electrode*

At threshold with steady potential depending somewhat on ratio of point diameter to gap length, positive streamers initiate at the surface and advance a short distance into the gap.[10] The field at some distance from the anode is adequate to create streamer-forming avalanches near the anode surface. These propagate only a short distance axially into the gas. When they cease to advance the positive-ion space charge they leave behind inhibits further streamers until it dissipates. Thus a succession of streamers follow as new triggering negative electron avalanches arrive at the anode and when the field has recovered sufficiently to permit streamers to form them. In consequence of this action, especially with more extensive electrodes, it is also possible for the photoionization to cause a spread of the discharge over the anode surface during the period when the positive space charge prohibits the streamer propagating axially from the surface.[10] These sequences of lateral discharges are called burst pulses. As ratios of gap length to anode radius and pressures vary, intermittent streamers or burst pulses dominate at threshold.[10] With potential increase above threshold, the streamers grow more frequent and burst pulses are temporarily prolonged. Space-charge inhibition increases as well. At some point the creation of negative ions by streamers and burst pulses is such that a negative-ion sheath forms near the anode surface, that is, between it and the positive space charge out in the gap. This zone extends outward a short distance from the anode surface where fields are inadequate to detach the electrons from the O_2^- created. This negative-ion sheath is known as the Hermstein sheath and yields the quiescent steady Hermstein[10] glow corona. The sheath produces a very high anode field but one that is too short to permit streamers to develop. Thus until the applied potential reaches more than 100 to 200% of the threshold value (depending on point radius to gap length ratio), streamers that can cross the gap and form a spark do not materialize. Eventually, the fields are so high as to permit streamers to form. These cross the gap and yield a spark.[10] If an impulse potential is supplied with a rise time on the order of 10^{-7} s or less, so that positive space charges and negative-ion accumulation in the gap do not form, the short streamers at threshold lengthen in proportion to the applied potential. That is, once the critical space charge leading to a self-propagating streamer tip forms (zero field streamer of Dawson and Winn)[11] it augments in strength as it falls through more and more of the decreasing field outward from the point. Thus in air it increases from a minimum 10^8 excess positive ions in a sphere of 3×10^{-3} cm radius to perhaps in excess of several times 10^9 ions in a sphere of perhaps 10^{-2} cm radius.[12] Such primary streamers can cross a gap of several centimetres branching heavily as they progress.[13] They leave behind a dark and relatively poorly conducting channel except near the anode. Primary branched streamers move with speeds ranging from 2×10^8 cm/s to 10^7 cm/s across the gap.[14] When the highly charged streamer tip reaches the cathode combining its high field at atmospheric pressure with its strong ultraviolet radiation, it sends a return stroke, or space wave of potential gradient, at about 5×10^8 cm/s back to the anode.[14] This ionizes and renders the channel more conducting. The wave interacting with the highly conducting channel region near the anode which results from convergence of electron currents from the branches in the high field region, enables this conducting channel, called a secondary streamer, to advance towards the cathode.[15] In longer gaps the secondary streamers may reflect the wave of potential gradient and send it back to the cathode. Having arrived at the cathode a new return stroke space wave then again moves to the secondary streamer tip and advances the secondary streamer. In very long gaps Kritzinger[7, 14] has noted at least five such potential gradient waves before the secondary streamer reaches the cathode essentially completing the arc channel. The general speed of advance of the secondary streamers lies in the 10^6 cm/s range.[14, 15] They are highly luminous and it is these that are usually photographed.

Recently Professor Oshige in the writer's laboratory has extended measurements for positive points from 200 torr down to 4 torr.[16] Here impulse potentials of somewhat longer duration could be used since negative-ion formation is not as rapid and the Hermstein sheath does not as readily develop. As pressures decrease the streamers still propagate from the anode. They become more diffuse and branching decreases materially at the lowest pressures. Speeds of propagation of the primary tips again lie generally in the 10^7 cm/s range. As pressures decrease the range of streamers rapidly increases so that gaps up to 19 cm are used at lower pressures in place of 4 cm at 760 torr. Near threshold primary streamer tip speeds are high near the anode. They decrease rapidly as the streamers advance towards the weaker fields at midgap. However at the lower pressures very active photons reach the cathode as the primary tip forms and advances. Electron avalanches from photoelectrons released advance towards the anode. Their flashes can be observed near the cathode by photomultiplier even before the streamer tip crosses. When these electron avalanches reach the retarded primary streamer tip in midgap, they enhance its speed and intensity so that the retarded tip can cross the gap near threshold. At somewhat higher potentials the speed remains fairly constant until it nears the cathode. Here the distortion of the field near the cathode by the streamer tip field again increases its speed. Once the primary tip crosses the gap its sends space waves of potential gradient sweeping up the main branches. Where the larger branch trunks converge near the anode but still in mid-gap, local regions of very high conductivity are produced. The resulting potential gradients directed towards anode and cathode accordingly initiate *highly conducting secondary streamers from midgap* that have *tips advancing to both the anode and cathode* at speeds of around 10^6 cm/s. The light emitted as these regions of high conductivity area created again gives rise to a heavy electron avalanche from the cathode. This avalanche can be followed by the photomultiplier in its advance towards the anode for some distance.

Once its tip appears to meet the downward moving positive tip of the secondary streamer the arc channel is complete and the arc develops.

B. *Highly stressed negative point*

Here the discharge starts at the cathode as a miniature glow discharge.[17] It is initiated by a photon secondary γ_p mechanism that builds up a positive ion space charge near the cathode. The electrons continue to ionize beyond this and in the lower field regions near the point create a negative-ion space charge by dissociatively attaching to molecules. The positive-ion space charge increases the cathode field and electrons soon gain their full ionizing energy in the very high field of the developing Crookes' dark space. The positive ions soon begin to reach the cathode in the high cathode fall field and yield a secondary γ_i action. However the negative O^- ion space charge created just beyond the positive-ion space charge by dissociative attachment weakens the field near the cathode and chokes off the discharge temporarily. The negative ions then dissipate towards the anode plane and a new discharge can build up. Thus at threshold and for a considerable potential range above this there is observed a relaxation oscillator-like pulsed discharge. The so-called Trichel pulses begin at about a thousand per second, and increase in frequency and decrease a bit in amplitude as potential increases up to a frequency of about 10^6 per second. At that point the cathode field is high and extends out so far that the O^- ions detach their electrons. In consequence the negative-ion space charge is created too far from the point to choke off the discharge and a quiescent pulseless glow discharge results. This continues for a certain range of potentials and is then replaced by negative streamers.[18] These streamers are just like those observed by Wagner in uniform fields. The range of potential for the Trichel pulse regimes, the pulseless regime, and the negative streamer regime varies rather widely as the ratio of point diameter to gap length and as pressure varies as Miyoshi has shown.[18]

The negative streamers are the same as the negative streamers noted in uniform field breakdown. Since they branch poorly, and are highly dissipative in character owing to the repulsion of the mobile electrons by the high streamer tip fields, they do not advance very far. They are confined to the high field region near the cathode alone. The larger the point radius and the shorter the gap, the more prominent they become. In fact they reach their peak efficiency in overvolted uniform field gaps. In any event with steady or even pulsed potentials of the order of microseconds a series of successive, more or less abortive, negative streamers start from the cathode.

At the potential at which they appear the negative-ion space charges with the larger corona currents augment the geometrical fields at the anode. In addition it appears that a highly polished anode plane of metal, or even of NaCl, will under the action of the negative ions cause the growth of crystallites or 'whiskers' of the anode material. Rautureau[19] and Goldman have shown by replicative electron microscopy that after a few minutes of negative corona discharge, crystals of a micron or so in diameter will grow in the anode surface. These whiskers, as Alpert *et al.*[20] have shown, can increase the anode field by factors as much as a hundredfold. They thus lead to positive streamers from the anode. These streamers increase in number with the growth and duration of the corona yielding the ion space charge. Thus as potentials approach spark-over with the negative point the negative streamers with their field distortion, the negative-ion space charge and the whiskers lead to *positive streamers* from the anode that advance towards the cathode. When the main branch tips of negative and positive streamers meet in midgap, space waves of potential gradient sweep to anode and cathode completing the arc channel. This action has been beautifully verified by Kritzinger[7] in photographs of interrupted long gap sparks and by photomultiplier studies. This sort of action was also earlier established in long spark photographs of secondary leader processes by Allibone and Meek.[21] The junction point of negative and positive streamers depends on the gap length and ratio of point radius to gap length at any pressure.

2. Free electron gases

A. Highly stressed positive electrode

In this case while negative-ion space charges play no role, the positive ions still exert inhibitory action as they accumulate. Thus one would expect that the long potential range between threshold and spark transition caused by the Hermstein sheath should be markedly reduced. As soon as streamers become vigorous enough to overcome space-charge effects of positive ions on application of steady potential they should cross and eventually yield a spark. With impulse potentials the *breakdown* spark transition should be still nearer corona breakdown.

Unfortunately relatively little work has been done on this aspect. The results obtained have been somewhat contradictory. This arises from the fact that aside from the arrival at a critical avalanche size the free electron gases must be capable of sufficiently active photoionization by their own photons to permit a streamer to build up. It is convenient in this respect to consider argon. Weissler[22] with so-called spectroscopically pure argon *observed no antecedent corona* with steady potential. Once the breakdown potential was reached there was a spark. Das[22] used argon under less rigorous purity conditions. He obtained results akin to those of Weissler when small amounts of contamination were present. That is, a streamer corona preceded spark breakdown which followed at about a kilovolt or more later. In argon purified by gas cataphoresis to remove traces of N_2 and Kr, argon did not yield a normal streamer breakdown.[5] The spark occurred as a rather hazy and somewhat broadened channel and the luminosity built up on a time scale of microseconds. It is probable that the streamer-like positive space charge that accumulated at the anode point was enabled to advance by the arrival of photoelectrons and the avalanches created by them from the cathode and that were liberated by the bright light of the local breakdown at the anode. All these electron avalanches need not have been liberated from the cathode but could well have come from less absorbed longer-range photoelectrons from the volume of the gas. Both H_2 and N_2 that were quite pure showed no burst pulses.[23] They gave localized coronas at the anode point at around 4000 volts for a 0·5 mm diameter point and a 3 cm gap.

These might have been incipient streamers in N_2. In H_2 there was just a glow at the anode. Breakdown streamers appeared at around 10 000 volts and the sparks ensued at around 18 000 volts. N_2 is known not to be readily ionized by its lower energy photons and the same applies to H_2. It takes much higher fields and electron energies to photoionize these gases and to produce space charge fields creating self-propagating photons. O_2 and especially N_2-O_2 mixtures are unique in their efficient photoionization. How far positive space charges play an inhibiting role on streamer advance in free electron gases is not known. In these studies, unlike the case for argon, photoelectrons from the cathode are unlikely at the high gas pressures owing to absorption of the active photons by the gas. In pure argon the photons from excited A_2 molecules are known to be effective as in Westberg's study. These are created in great profusion at near 400 torr as shown by Colli and Facchini.[24]

More work must be done in studying the breakdown in different free electron gases from positive points with impulse potentials in pure gases, especially using two photomultipliers and modern techniques.

B. Highly stressed negative electrode

The mechanism in this case has been recently clarified by Hassoun[25] in the writer's laboratory. The writer[26] in 1949 had indicated the inherent instability of the highly stressed negative electrode in free electron gases in consequence of studies of Weissler,[22] Miller,[27] and on the basis of the action of the positive-ion space charge near the cathode surface. This breakdown was investigated in spectroscopic grade He with Alpert vacuum techniques using a flashed hairpin filament tungsten point in a point-to-plane gap of 4 cm length. Flashing the cathode was essential to prevent liberation of gases through the sputtering of the cathode by ion bombardment. Pressures ranged from 50 to 220 torr. Current-potential, current-photomultiplier, and photomultiplier-potential observations of luminous events were recorded. Two photomultipliers were used, one at the cathode, the other at various points in the gap which noted the progress of luminosity across the gap. These photomultiplier techniques were used with 3889 Å and 5876 Å filters corresponding to two He lines, one of which (5876 Å) characterized electron impact excited levels and the other (3889 Å) was more characteristic of the arc phase. Still photographs enabled the shape and size of the arc channel at current peak to be observed. Impulse potentials of short rise time, 10^{-8} s, from a storage condenser, were used. Fortunately the long statistical time lag owing to the confined high field area near the point allowed steady potentials to be arrived at before the breakdown began. Thus the breakdown could be studied at various potentials above threshold. By varying the capacity which was discharged the duration of the arc could be controlled. If potential and capacity were great enough the transient cold cathode arc by electrode heating changed to a hot cathode arc. This phase was not investigated.

The discharge initiated as a Townsend discharge with a photoelectric secondary γ_p. As indicated under the Trichel pulse analysis it quickly built up a space charge with high cathode field followed by the subsequent secondary liberation by energetic positive-ion bombardment. Negative ions did not form. Thus the discharge was not interrupted as in air. However the photomultiplier oscillograms indicated oscillations of 7×10^7 cycles/s caused by ion plasma oscillations with a positive-ion density of $4 \cdot 5 \times 10^{11}$ electrons/cm^3. These were constant during the duration of the luminosity and on occasion still photographs of the arc showed striations. These oscillations were independent of instrument and circuit constants and were an inherent property of the arc. It is believed that with the observed exponential increase with time, the electrons accumulated beyond the ionic space charge more rapidly than they were removed by the field. Thus even here the electron space charge was sufficient to reduce the current momentarily, allowing the discharge to spread laterally over the electrode surface and setting the ionic space charge into oscillations characteristic of the ion density. Similar oscillations have been observed and theoretically predicted in the case of a positive coaxial cylindrical discharge with positive wire in argon by Colli, Facchini, Gatti, and Persano.[28] As the negative electron space charge develops near the cathode with exponentially increasing current it builds up a gradient front with an exponential tip profile that propagates across the gap at the speed expected of electrons in the field on the anode side of the gap. This ranges from 4×10^6 cm/s at the cathode to 10^6 cm/s near the anode. The gradient front is steep enough to excite and ionize the gas as it advances. Thus a transient luminous pulse with an exponential rise advances across the gap. As observed with the filter sensitive to light by direct electron excitation (5876 Å), this light pulse viewed at any one point rises somewhat more rapidly than the exponential rise of the current and declines before the current begins to decline. The current peak on the other hand is reached by the time the potential on the source has fallen to about half value. The luminosity with the arc (3889 Å) filter peaks with the current but persists well after current appears to have declined. This action would be expected of a quasi-thermal arc. The duration of the excitation light pulse (5876 Å) is such that the width of this luminous excitation phase, taking into account the speed of advance is about the order of the gap length.

The data taken using the width of the photographed arc channel permit estimates of the current density at current peak. This appears to be nearly constant at different potentials. The potential across the arc is also constant. From these data and the estimated temperature of the arc derived from the diffusion during rise to peak to yield the arc channel diameter, it is estimated that the quasi-thermal arc is established well above threshold. The arc temperature then lies between 2000°K and 3000°K at 2150 and 2400 volts applied across the paths but with electron densities of the order of 1×10^{14} ions/cm^3. The ion density in the current channel is much larger than the $4 \cdot 5 \times 10^{11}$ ions/cm^3 in the ion sheath responsible for the ionization at the cathode determined by the plasma oscillation frequency. This discrepancy is, however, readily understood when one notes that the current of the arc is confined to a channel created by the initial ionized path across the gap that can only spread radially by electron diffusion and is later contained as an ion plasma channel. At current peak the whole area of the cathode is involved in generating the current funnelling down into the confined channel. Estimates of the relative areas of cross-section of the arc channel to luminous cathode surface are in rough agreement with the difference in charge densities.

It is seen here that the breakdown is by an entirely different mechanism from that for the electron attaching gases and for those from the highly stressed positive point in photoionizable gases. It is *not* a streamer mechanism. This mechanism is probably much the same as that for the breakdown of the lower pressure glow discharges from disc electrodes. There the discharges probably start from the sharp edges of the cathode and as streamers from the sharp edges of the anode. As potentials increase these merge into a single channel. This action may be seen in the breakdown of such a Geissler discharge tube as the pressure is being lowered with the high potential applied.

References

1. Raether, H., *Electron Avalanches and Breakdown in Gases*, Butterworths, London, (1964).
2. Tholl, H., *Z. Naturforsch* **19a**, 346, 704, (1964).
3. Wagner, K. H., *Z. Physik* **189**, 465, (1966).
4. Przybylsky, A., *Z. Physik* **151**, 264, (1958); *Z. Naturforsch.* **16a**, 1232, (1961); Sroka, W., *Physics Letters* **14**, 301, (1965); Teich, T. H., *Z. Physik* **199**, 378, (1967).
5. Loeb, L. B., Westberg, R. G. and Huang, H. C., *Phys. Rev.* **123**, 43, (1961).
6. Loeb, L. B., *Science* **148**, 1417, (1965); Winn, W. P., *J. Geophys. Res.* **70**, 3265, (1965); *J. Appl. Phys.* **38**, 783, (1967).
7. Kritzinger, J. J., *The Breakdown Mechanism of Long Sparks in Air*, Ph. D. Thesis, Dept. Electrical Engineering, University of Witwatersrand, Johannesburg, (Jan., 1962).
8. Köhrmann, W., *Z. Angew. Phys.* **7**, 183, (1955), *Ann. Physik* **18**, 379, (1956); Schroeder, G. A., *Z. Angew. Physik* **13**, 296, (1961).
9. Shale, C. C., *Physical Phenomena Underlying Negative and Positive Coronas in Air at High Temperatures and Pressures*, 1965 International Convention Record IEEE, Part I, Chap. II, Sec. G. p. 50 ff.
10. Loeb, L. B., *Electrical Coronas*, University of California Press, Berkeley, (1965), Chap. 3, p. 75 ff., Hermstein Glow Regime, p. 86 ff.
11. Dawson, G. A. and Winn, W. P., *Z. Physik* **183**, 159, (1965).
12. Ref. 10, Appendix II and IV.
13. Ref. 10, Lichtenberg figure studies of Nasser, pp. 168-221.
14. Ref. 10, Appendix I, p. 619, Appendix II, p. 631.
15. Hudson, G. G. and Loeb, B., *Phys. Rev.* **123**, 79, (1961).
16. Oshige, T., *J. Appl. Phys.* **38**, 2528, (1967).
17. Ref. 10, Chap. IV, pp. 299-311, pp. 318-321, pp. 335-347, pp. 383-402; Chap. 6, pp. 513-551.
18. Ref. 10, pp. 372-382; work of Miyoshi, p. 383 ff; Appendix II, work of Kritzinger, p. 624.

19. Tchoubar, M., Goldman, M., Rautureau, M. and Buchet, J., *Microscopie* 3, 511, (1964); Rautureau, M., *Diplome Thesis,* Faculté des Sciences, Univ. of Paris, (June 1965).
20. Alpert, D., Lee, D. A., Lyman, E. M. and Tomaschke, H. E., *J. Vac. Sci. and Technol.* 1, 35, (1964).
21. Allibone, T. E. and Meek, J. M., *Proc. Roy. Soc.* **A166**, 97, (1938), **A169**, 246, (1938).
22. Ref. 10, Weissler's studies in Ar and Das' studies, p. 366, 368.
23. Ref. 10, p. 362 ff.
24. Ref. 10, Colli and Facchini's work, p. 425.
25. Hassoun, A. M., *The Negative Point-to-Plane Breakdown in a Free Electron Gas,* Doctoral Dissertation, University of California, June 1967. Manuscript submitted to *J. Appl. Phys.*, (Sept. 1967).
26. Loeb, L. B., *Phys. Rev.* **76**, 255, (1949).
27. Ref. 10, 515 ff.
28. Colli, L., Facchini, U., Gatti, E. and Persano, A., *J. Appl. Phys.* **25**, 429, (1954).

Introduction to Paper 20 (Stekolnikov and Shkilyov, 'The development of a long spark and lightning')

The electrical breakdown of long discharge gaps is of considerable interest to high-voltage engineers and becomes of greater importance as the voltage at which electrical power is transmitted is increased. The discharge gaps of greatest interest technologically are those giving non-uniform electric fields, e.g. rod-plane, sphere-plane, and rod-rod gaps. Experimental studies have usually used rod-plane geometries with gap separations of from 10 cm to some metres in length. The voltages used are impulse voltages and the exact shapes of the pulse waveforms, degree of overvoltage, presence of space-charge from corona discharges and so on have been found to be of importance in the breakdown of the discharge gaps. The study of long discharge gaps in the laboratory is also attractive from the point of view of understanding the mechanism of the lightning discharge, the most dramatic and uncontrollable of all breakdown phenomena. The two areas of interest merge when one considers the problems of protecting electrical equipment from lightning strokes.

A number of quite different techniques have been used to study the breakdown of long discharge gaps, some of which are described in the material suggested below for further reading. Optical techniques have been particularly popular and one of the more successful has been the use, as in this paper of Stekolnikov, of image-converter cameras (or electron-optical converters as they are referred to by Stekolnikov and others).

The paper given here is a good example of the excellent work of Stekolnikov and his co-workers. It was given at a conference on atmospheric and space electricity and is particularly relevant to theories of lightning discharges. It is, of course, extremely difficult to explain quantitatively the kind of data obtained in this and similar investigations and to apply the results to lightning.

Further Reading

Meek, J. M. and Craggs, J. D., *Electrical Breakdown of Gases*, Oxford University Press, 2nd edition, to be published.

Saxe, R. F. and Meek, J. M., *Proc. I.E.E.*, (1955).

Park, J. H. and Cones, H. N., *J. Res. Nat. Bur. Stand.* **56**, 201, (1956).

Waters, R. T. and Jones, R. E., *Phil. Trans. Roy. Soc.* **206**, 185, (1964).

Kritzinger, J. J., appendix I of L. B. Loeb's *Electrical Coronas*, University of California Press, (1965).

20

The development of a long spark and lightning

I. S. STEKOLNIKOV AND A. V. SHKILYOV*

The data on a long spark are used when analysing the mechanism of lightning (which, in most cases, has a negative polarity), when estimating the quantitative characteristics of its stages, when studying protection areas of lightning diverters and the like. However, until recently, a spark was optically investigated by the Boys camera, which is not perfect in this respect because of its limited scanning speed and poor light sensitivity. On the other hand, the often-used method of 'chopping' of separate spark stages for obtaining successive series of static photographs of its development did not always have sufficient light sensitivity. In this connection, some investigators made serious mistakes when interpreting pictures of spark development as far as the structure of the impulse corona 'element', the velocity of the impulse corona development, the succession of the spark stages in time (for instance, in the −rod +plane gap), or other similar questions were concerned.

To obtain more detailed data on the long spark, its development was studied in the +rod −plane, the −rod +plane, and the −rod +rod/plane gaps. The investigation was carried out by an electron-optical converter (E.O.C.) with light intensification and by electron high-speed oscillographs. The time-scanning speed of the E.O.C. can attain several 10 000 km/s and its high optical sensitivity permits distinct fixation of poor light fluxes of the spark pre-discharge stages. Detailed study of these discharge processes was not accessible with any of the previously used investigation equipment.

The study of the negative spark on the $1 \cdot 5/1000$ μs wave

This study was carried out in the −rod +plane and the −rod +rod/plane gaps (Stekolnikov and Shkilyov, 1962a, b). A 20 mm rod with a hemisphere at one end served as a negative electrode; a 10 mm rod with a hemisphere at one end, the height of which was $2 \cdot 5$–50 cm, was set on the plane in the case of the −rod +rod/plane gap. The gap length S_0, in the case of −rod +plane, was 100–300 cm; in the case

* Reproduced from *Problems of Atmospheric and Space Electricity*, edited by S. C. Coroniti, Elsevier, Amsterdam, (1965) by permission of The Elsevier Publishing Company, Amsterdam.

of −rod +rod/plane, S_0 equalled 270–300 cm. The voltage wave was $1 \cdot 5/1000 \, \mu$s and had an amplitude U close to the minimum discharge amplitude ($k = U/U_{min} \approx 1 \cdot 0$). The E.O.C. with the following lenses was used: quartz ($f/3 \cdot 5$); 'Jupiter-3' ($f/1 \cdot 5$); and 'Jupiter-12' ($f/2 \cdot 8$). The E.O.C. electrostatic shutter worked on the 'open–close' principle so that a static picture of the discharge processes already developed at the beginning of time scanning was obtained. The synchronization of the E.O.C. and oscillograph records was provided by connecting their time plates with one scanning voltage source. The current was recorded with a shunt connected with a measuring plane of 3 x 3 m. The latter was placed at the height of 15 cm from the earthed plane of 8 x 8 m.

For the purpose of more detailed analysis of the time-scanning picture of the spark development obtained on the 'eocograms', the discharge processes were occasionally simultaneously photographed by a stationary camera with a quartz lens ($f/4 \cdot 5$). To reduce the light flux from the main channel and the arc, the 'chopping' of voltage in the gap S_0 by another gap $S_1 < S_0$ was used.

Figure 1 gives the diagram of the spark development in the −rod +plane gap. The negative impulse corona forms a region of negative space charge; if U_{min} is chosen, the radius of the region does not exceed $\frac{1}{3}S_0$. On the region boundary of

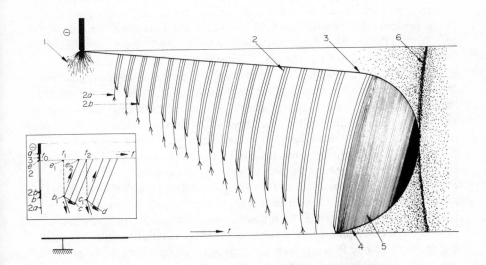

Figure 1. The diagram of the spark development in the −rod +plane gap. 1 = the impulse corona; 2 = the stepped leader; $2a$ = the branches of the impulse corona; $2b$ = the step (the stem of the impulse corona); 3 = the negative leader; 4 = the positive leader; 5 = thread-like luminosity (the leader corona); 6 = the main channel.

The diagram in the insert shows the development of the process when the next lengthening of the stepped leader occurs (a, e and b indicate the static picture preceding the time scanning; e_2, b_1 and c the time scanning).

the negative space charge, certain conditions are created so that it could propagate in one or two directions towards the plane in the form of the stepped leaders of the spark. The effective speed of propagation of these stepped leaders varies within the range of 0.8×10^7-2×10^7 cm/s. When the next lengthening of this leader occurs, three processes take place (*Figure 1*): the impulse corona branch propagates towards the plane at a velocity of 1×10^8-3×10^8 cm/s in the form of one or more filaments 25–50 cm long; the tip of the step (which is evidently a stem of the impulse corona) also propagates towards the plane at a velocity of the order of 5×10^7 cm/s (full lengths of the steps are given from 5–15 cm); and the luminosity wave (the brightness of which is more intensive in lower portions) advances up the stepped leader channel which was formed before at a velocity of about 10^8 cm/s. During the step development time, such luminosity waves may be observed to repeat several times and they are always directed upwards from the moving tip of the step. There are pauses on the order of 1 μs between the leader steps. The end of the preceding step is usually the beginning of the next step.

The mechanism of the stepped leader shown in *Figure 1* can be qualitatively described in the first approximation in the following way. The impulse corona which flashes from the rod impinges the negative space charge into the discharge gap. Most of the density of this charge is at the border of the branches. Here, gradients reach the critical value which gives rise to the development of new discharge processes, i.e., impulse corona flashes occur which consist of a stem and branches. As the negative charge is removed to the boundary region of the new impulse corona flash, the positive space charge is exposed in the region of its appearance. In this connection, the electrons from the near-electrode region flow towards it; this again increases gradients in the region of the negative charge which has been produced by them. Thus, impinging of the negative charge in the form of the stepped leader channel of the spark begins. The next lengthening of the stepped leader channel of the spark takes place at the critical gradients of the field from the point b_1 (see *Figure 1*) in the region of its tip. The impulse corona flashes. As the negative charge is removed from the region b_1, the positive space charge is exposed here. Therefore, electrons flow along the stepped leader channel from the electrode towards the positive space charge. This gives rise to ionization and the luminosity which accompanies it propagates as a wave from the point b_1 to the point e_2. The development of few successive flashes of ionization along the path b–e can be caused by fluctuations in the process of neutralization of the positive space charge in the region b_1, either by the electrons flowing into that region or leaving there. The removal can be done while the repeated impulse corona branches (sometimes invisible) are being formed, including those from the stem. The next step of the leader appears after the stem of the first step impulse corona has developed. The presence of a relatively narrow channel full of negative charges will cause sufficient gradients for initiation at the point c_1 (*Figure 1*) of a new impulse corona flash which consists of branches and a stem. The process of formation of the next stepped leader stage of the spark repeats. The stepped leaders cause the air to ionize and advance a negative charge deep into the gap. This negative charge changes the original distribution of gradients along the gap: the gradients increase up to the critical value near the grounded objects.

The stepped leader of the spark which is described here differs considerably from those described by Schonland *et al.* (1935), Griscom (1958) and Wagner and Hileman (1961). Simultaneously with the propagation of the stepped leader towards the plane, an ordinary negative leader is developed in its channel at a velocity of 1×10^6–5×10^6 cm/s from the rod. At the moment when the stepped leader channel touches the plane, the positive leader begins developing from the plane; the positive and negative leader coronas connect which permits the negative charge to flow out of the gap to the plane. This, in its turn, contributes to the sharp increase in the leader velocity. The field between the tips of the approaching leaders becomes uniform which produces a final jump. In the −rod +rod/plane gap (*Figure 2*), the stepped leader of the spark gives rise to the development of the opposite process in the form of positive impulse corona branches from the grounded rod $h/S_0 = 0\cdot01$–$0\cdot2$ high; the mean velocity of their tips is about 5×10^7 cm/s; the process arises at a certain critical distance S_{cr} of the stepped leader channel from the plane. The velocity of the branch tips increases as the branches approach the stepped leader. At the moment when the branches and the stepped leader contact, luminosity blazes up along the whole gap and the positive leader starts developing from the rod. S_{cr} depends essentially on h of the rod and its displacement about the gap axis. The decreasing of h causes S_{cr} to decrease. At last, if the plane is smooth ($h = 0$), the stepped leader touches the plane and, from

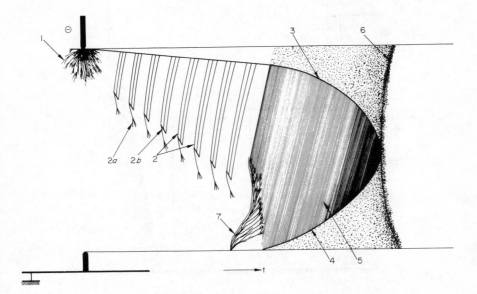

Figure 2. The diagram of the spark development in the −rod +rod/plane gap. 7 = branches of the positive impulse corona; the rest of the symbols are the same as in *Figure 1*.

the point of their contact along the stepped leader path, an ordinary opposite positive leader develops at an initial velocity of about 10^7 cm/s. When the positive leader appears, the velocities of the positive ordinary leader and the negative ordinary leader become approximately equal and rise rapidly for both the −rod +plane gap and the −rod +rod/plane gap. A sharp increase of the tip velocity can occur in any section of the gap while the leader is developing. In the time-scanning pictures, this increase of the tip velocity appears to be a jump-like lengthening of the channel, and this is called a 'jump' (Stekolnikov and Shkilyov, 1960). The jump duration is on the order of 10^{-8} s and the length is 15–125 cm at $S_0 = 500$ cm in the +rod −plane gap. While examining the oscillograms of current, it was found that the leader is degraded (Stekolnikov, 1957). This phenomenon is due to the fact that the field in the discharge gap becomes uniform. Ionization processes take place almost simultaneously along the residual gap; this, perhaps, accounts for the rise of jumps. The jump with which the leader stage is concluded is called final. Perhaps this phenomenon is similar to a 'pre-strike' in lightning (Griscom, 1958).

The diagram plotted in *Figure 3* was brought forward by Schonland and his associates on the basis of photography of lightning by the Boys camera. Captions in the figures define the element of this diagram. Comparison of the diagrams in *Figures 1* and *3* shows that they are far from being similar.

As mentioned, if in the case of a spark, no main channel and no breakdown are formed after the stepped leader touches the plane, lightning which strikes the plane

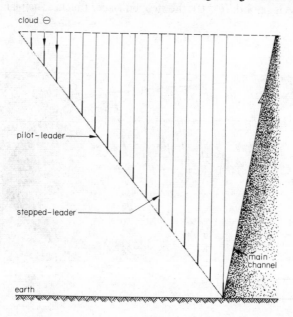

Figure 3. The diagram of lightning leader development (Schonland *et al.*, 1935; Schonland, 1938, 1953).

is assumed to develop the main channel immediately after the stepped leader approaches the ground surface; the stepped leader path is considered to be highly conducting. Further lengthening of the stepped leader of the spark does not correspond to the described mechanism of the lightning leader step. Our pictures, taken by apparatus of high optical sensitivity, did not show the 'pilot-leader' of the kind postulated by Schonland (1938, 1953).

Taking the obtained data into consideration, it is possible to assume that the diagrams of lightning development based on Boys pictures must be reconsidered. We conclude that the lightning development must be studied with E.O.C. The diagram of the spark development in the −rod +rod/plane gap (*Figure 2*) differs essentially from the assumed diagrams which show the lightning stroke to the metal mast (the lightning arrestor). This case is of great importance for calculations of overvoltages which are produced by lightning striking a tower of a transmission line.

The investigation of the spark on exponential waves

Tests which had been carried out in our laboratory show that the strength of the discharge gap (+rod −plane) greatly depends on the shape of the wavefront (Bazelyan *et al.*, 1960; Stekolnikov *et al.*, 1962). On the exponential wavefront (conventionally called an 'oblique wavefront'), the voltage-time characteristic for the +rod −plane gaps has a U-type form with the minimum V_d within the range of 150–300 μs (*Figure 4*). Under this condition for the rod −plane gap length $S_0 = 4.5$ m, the average discharge gradient $E_{av} = 2.4$ kV/cm instead of $E_{av} = 5.35$ kV/cm on the standard wavefront.

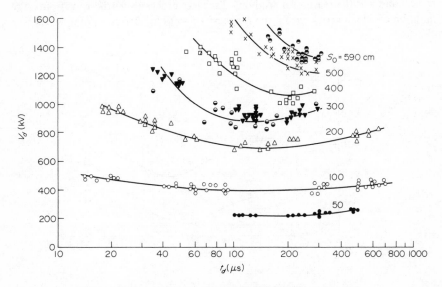

Figure 4. Voltage-time characteristics of the +rod −plane gaps on exponential (oblique) waves of voltage.

As a result of these experiments, it was assumed that the space charge, which is impinged first by the impulse corona and then by the leader channel and its corona, is a main factor which causes large E_{av} in the left-hand side of the U-type characteristic (*Figure 4*). The charge which is impinged by the impulse corona is, evidently, proportional to some extent to its intensity and to the branch velocity. For instance, it has been found that for the +rod –plane gap of S_0 = 200 cm, at the average steepness of voltage front A = 500 kV/µs and the 2/1000 µs wave, the effective velocity of the impulse corona ($v_{i.c.}$) was 2×10^7 cm/s. If A is 1·7 times more than that in the previous case and the waveshape is the same, $v_{i.c.}$ will be 10^8 cm/s. As A is lowered, the charge which is impinged into the gap is also reduced. The increase of the discharge voltage in the right-hand side of the U-type characteristic can be caused by accumulation of the space charge in the gap which has been impinged, as it is shown in the time-scanning photographs, by numerous flashes of the corona stems and by the leader corona. The study of the discharge mechanism within various ranges of voltage-time characteristic by the E.O.C. has shown that V_d depends on the length to which the unbridged gap is shortened at the moment the ionization region, in the form of the impulse corona and the leader corona, touches the plane. This, in its turn, makes it possible to remove the space charge from the gap, which accelerates the development of the leader and completion of the breakdown if voltage is still applied to the gap.

Recently, the U-type voltage-time characteristic was found at the negative voltage for S_0 = 1·0 and 1·5 m (*Figure 5*). It should be noted that if $t_d > 20$ µs, a considerable scattering of values of V_d and t_d is observed for each of the fixed waves. This characteristic is of interest both in the practical and in the theoretical respect and was the reason for studying discharge processes on the wavefronts, the duration of which varies within the range of 1·5–400 µs. Corresponding waves were

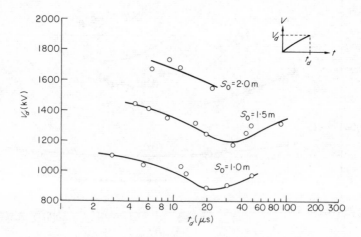

Figure 5. Voltage-time characteristics of the −rod +plane gaps on exponential (oblique) waves of voltage.

formed by front-building capacitance, C_f = 1000 pF, connected in parallel with the discharge gap; front-building capacitance was charged by the impulse generator (C_{output} = 0·02 x 10^{-6} F) through the resistance R_f, which varied within the range of 2·5 − 350 kΩ. Eocograms and oscillograms obtained gave new data on the spark development in the −rod +plane gap.

The fact that the exponential wave is applied to the discharge gap means that the steepness of voltage front is continuously changing. However, this variable steepness of voltage front can be approximately characterized by a certain constant average steepness (rate) A; for instance, that between t = 0 and t = 2·3R_fC_f. Thus, the exponential wave offers a good way of affecting the discharge gap with waves of various average steepness of voltage front; but the change of the steepness during the voltage-front time does not considerably affect discharge processes if R_fC_f remains constant. The decrease of A to 100-250 kV/μs, as compared with the standard wave where A = 600-100 kV/μs, causes discharge processes to change. *Figure 6A* shows that the impulse corona appearing at the beginning of the discharge was not over 25 cm in length. The stepped leader of the spark is preceded by an impulse corona. In this case the voltage-wave amplitude (*Figure 6B*) is somewhat less than V_d, and A = 160 kV/μs until the time t_1, the effective velocity of the stepped leader is 1·2 x 10^7 cm/s; the velocity of the negative leader, which propagates along the stepped leader path, is about 1·4 x 10^6 cm/s. At the next interval, $t_2 - t_1$, A_{av} = 40 kV/μs, and after t_2, the rate of voltage rise decreases still more. The stepped leader stops propagating because of the low voltage amplitude. *Figure 6C* illustrates the process of further stepped leader development where the voltage amplitude U was high enough to cause breakdown. The start of the time scanning was at the moment t_0', when the stepped leader had advanced 125 cm into the gap. Several stepped leader channels, near which one can see the impulse corona branches which took place during the stepped leader propagation, are distinguished on the static picture preceding the start of time scanning. Lengthening of the channel *a–b* alone is visible after the moment t_0'. The branch of the impulse corona appearing at the moment t_1' from the point b_1 reaches the plane. Simultaneously, the waves of luminosity described above propagate in consecutive order up the stepped leader path at a velocity of 1·3 x 10^8 cm/s. By t_2, the stepped leader channel grows towards the plane; the positive leader develops from the plane at an average velocity of 1·2 x 10^7 cm/s; now the luminosity extends along the whole gap. This thread-like luminosity (filaments which are probably traces of the flowing spheres of the kind which have been described by Park and Cones (1956) and which propagate at a velocity on the order of 10^8 cm/s) connects the tips of the positive and negative leader. The E.O.C. shutter was closed at t_3'.

The 'lacing' was found (Stekolnikov and Shkilyov, 1962b) to develop in the stepped leader channel of the spark. These lacings are also observed to develop on exponential waves. In *Figure 7*, the lacing develops from the moment t_1 in the advancing stepped leader channel from the point P in the gap space towards the points m and n. The development of this type of a channel permits calling it a

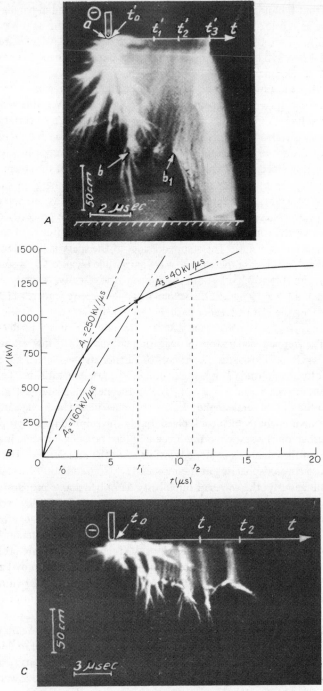

Figure 6. Discharge time-scanning picture (*A*, *C*) in the −rod +plane gap, S_0 = 2 m in length, (*B*) gives the shape of voltage wave.

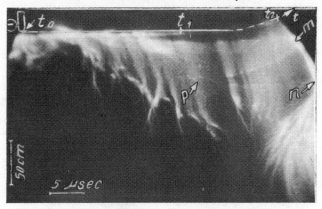

Figure 7. Eocogram of the stepped leader with the space leader
developing in its channel *P-m-n* (−rod +plane, S_0 = 2 m;
distortion of the scanning line t_0-t_1-t_2 is due to the fault of the
E.O.C.).

'space leader'. The channel develops in both directions from *P* (*Figure 7*). The
velocity of the positive tip of the space leader is on the order of 3×10^6 cm/s
and that of the negative tip is 0.3×10^6 cm/s. The velocity of the negative leader
tip (from the electrode) is about 0.3×10^6 cm/s. At the moment t_2, the positive
tip of the space leader touches the negative leader at the point *m*. The impulse
corona is formed simultaneously at point *n* from the lower tip of the space leader;
the fact that the impulse corona is formed at the moment of joining of the space
leader with the negative leader permits the assumption that the channel of the
space leader conducts very well and its potential before touching differs considerably from that of an electrode. Therefore, the potential of the negative tip of the
space leader increases sharply at the moment of joining. The negative leader channel
is lengthened by jumps of the length of the space leader as a result of those processes
described.

If $A < 40$ kV/µs, the stepped leader does not develop as an uninterrupted process
from the rod to the plane. At first, successive flashes of the impulse corona with
simultaneous lengthening of the leader channel from the rod occur near the rod at
some intervals (*Figure 8*). Near the rod, the impulse corona produces a region of
space charge as found on the wavefront of 1.5 µs (Stekolnikov and Shkilyov,
1962b); however, this region is formed for a longer period of time by several
separate flashes of the impulse corona. At the same time, the leader from the rod
is lengthened. Thus, from a particular moment, the development of the spark will
be similar to that on the standard wave. The development of this process until the
impulse breakdown stage is shown in *Figure 9A*; the start of the time scanning is
made at the moment t_0, before which several flashes of the impulse corona with

Figure 8. Push-like propagation of the negative leader (−rod +plane, S_0 = 2 m).

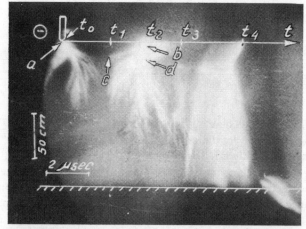

A

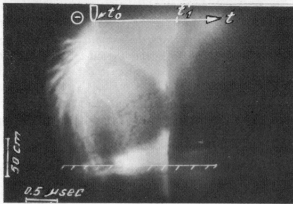

B

Figure 9. (*A*) The eocogram shows the stage preceding the final stage of discharge (S_0 = 1·7 m). (*B*) Eocogram of the positive leader and the negative leader propagating towards one another (S_0 = 1·9 m).

lengthening of the leader channel were formed in the gap. These elements, un-scanned, are seen at the beginning of the eocogram. The space leader begins develop-ing from point C at the moment t_1; it touches the electrode channel at point b at the moment t_2. The potential of the negative tip of the space leader at point d becomes equal to the electrode potential (this process has the form of a jump) and an impulse corona develops from the tip. Though it does not reach the plane here, it creates conditions for further development of the stepped leader of the spark which, advancing to the plane, gives rise to an opposite leader; this, in its turn, stimulates the development of the negative leader. The last stages of this process can be seen in *Figure 9B*. Here, the positive leader and the negative leader develop at about the same increasing velocity. The E.O.C. shutter closed before the leader channels touched.

The data obtained make it possible to plot principle diagrams (*Figure 10*) of the negative spark development in the −rod +plane gap ($S_0 = 1\cdot5$-$2\cdot0$ m). The diagrams correspond to certain ranges of varying A of the voltage wave which was applied to the gap. The diagrams are the average picture of spark development composed of a large quantity of eocograms. *Figure 10A* shows the spark development at $A = 250$-100 kV/μs. At the critical value of voltage V_{cr}, an initial impulse corona, the branches of which are not very long, develops from the rod. Then, as V increases, new elements of the impulse corona are produced (a repeated impulse corona). Both types of impulse corona produce the region of space charge near the rod; but this region is much less than in the case of the wave $1\cdot5/1000$ μs. Further increasing of voltage V gives rise to the development of the stepped leader, along the channel of which the negative leader develops from the rod. The phenomena described above are accompanied by current impulses measured at the plane; they characterize space charges impinged by the spark elements into the gap. After the stepped leader channel touches the plane, the positive leader arises from the latter (Stekolnikov and Shkilyov, 1962a, b). A speedily increasing current accompanies the development of leaders. As the tips of both leaders are connected by threadlike luminosity, ionization takes place along the whole gap; it extends from the positive leader to the negative leader in the form of luminosity waves and the whole gap becomes conductive. This causes the leader tip velocities to become equal and the leader lengths to increase rapidly. The gradients between the leader tips become uniform and grow with the approach of the leaders until the final jump occurs (Stekolnikov and Shkilyov, 1960). The effective velocity of the stepped leader is $(0\cdot7$-$1\cdot2) \times 10^7$ cm/s. The velocity of the negative leader (before the appearance of the positive leader) is $(0\cdot8$-$2\cdot1) \times 10^6$ cm/s. The initial velocity of the positive leader is on the order of 10^7 cm/s. Thus, if $A = 250$-100 kV/μs, the breakdown has the following stages: (*1*) the initial impulse corona and the repeated impulse corona, (*2*) the stepped leader, (*3*) the negative leader, which develops simultaneously with the stepped leader, (*4*) the positive leader, which is connected with the negative leader through the corona, (*5*) the final jump, and (*6*) the main channel.

If $A = 100$-50 kV/μs (*Figure 10B*), the beginning of the discharge is similar to that in the diagram of *Figure 10A*. However, when the stepped leader is at a

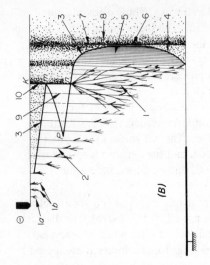

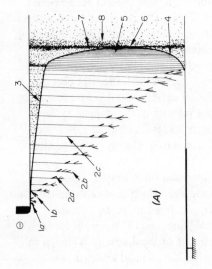

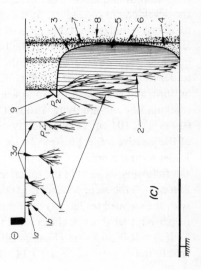

Figure 10. The diagram of the negative spark development in the −rod +plane gap on exponential (oblique) waves of voltage.

(A) At $A = 250–100$ kV/μs. $1a$ = the initial impulse corona; $1b$ = the repeated impulse corona; 2 = the stepped leader; $2a$ = the step/stem of the impulse corona; $2b$ = the branches of the impulse corona; $2c$ = the channel; 3 = the negative leader; 4 = the positive leader; 5 = thread-like luminosity (the leader corona); 6 = the final jump; 7 = the main channel; 8 = the arc.

(B) At $A = 100–50$ kV/μs. 1 = the impulse corona; 9 = the space leader; 10 = the leader channel flash; the rest of the symbols are the same as for diagram A.

(C) At $A < $ kV/μs. $3a$ = the push-like negative leader; the rest of the symbols are the same as for diagrams A and B.

considerable distance from the plane, the 'space leader' is formed and develops along the stepped leader channel. The space leader has two tips; the upper positive tip moves towards the negative leader (in the direction of the rod) and the negative one develops towards the plane. To account for the development of the space leader, it is assumed that the positive space charge is formed below point P when the impulse corona, which removes the negative charge, develops. If the stem of the preceding step is assumed to keep rather high conductivity, it may serve as a point from which the space leader channel will develop to the field created by the negative leader and by the positive charge; the latter arises periodically in the stepped leader development time. The tips of this channel propagate along the paths P-m and P-n. At the moment when the tip m of this channel and the negative leader meet, the length of the leader 3 increases by a section m-n with high conductivity, and the point n rapidly assumes the electrode potential. This process is similar to application of the wave potential with the high rate of voltage rise to the 'electrode' which is moved into the gap. If the instantaneous voltage in the gap is sufficient for the gap S_1 ($S_0 - k$-n in length) to be bridged, the breakdown occurs after the scheme which is given for the standard wave of voltage. As in *Figure 10C*, if the space leader propagating from the point P_1 is not able to create necessary conditions, no breakdown occurs.

The velocity of the negative tip of the space leader is close to the negative leader velocity, and the positive tip velocity exceeds it. In this stage, the space leader exists as a part of the stepped leader channel. After the negative leader is connected with the space leader, the potential of the space leader's lower tip, which becomes the tip of the negative leader from that moment on, rises rapidly enough to form the intensive impulse corona.

In cases where the branches of the impulse corona do not reach the plane, the stepped leader develops. The fact that the latter touches the plane gives rise to the positive leader and further, the leaders join. If the branches of this corona reach the plane, the positive leader develops with no preceding stepped leader.

Thus, if $A = 100$-50 kV/μs, the breakdown has the following stages: (*1*) the initial impulse corona and the repeated impulse corona, (*2*) the stepped leader, (*3*) the negative leader, (*4*) the space leader, its joining with the negative one, and (*5*) the breakdown similar to that on the standard wave.

It should be noted that a velocity on the order of 5×10^6 cm/s is critical for the stepped leader and it probably cannot develop at a lower velocity. The space leaders (several space leaders can develop in a stepped leader channel) are usually observed to form within 10 μs after the moment the first discharge processes start. Joining the negative leader channel, the space leader causes the unbridged gap length to decrease sharply and, therefore, the reduction of the discharge voltage V_d is observed on the oblique wave.

If $A = 40$-3 kV/μs, the discharge process begins with the initial impulse corona and the repeated impulse corona (*Figure 10C*). Simultaneously, a channel of the kind of a leader is formed; the new impulse corona develops from the end of this. This process repeats in some pauses, and the leader channel and the branches of the impulse corona increase in length every time. A phenomenon of this kind was observed by Walter (1899) and he called it a 'push-like leader'. The reason for such development of the process is probably because the space leader which joins the channel developing from the rod brings about a rapid increase of the potential of its tip; due to this fact, the impulse corona is formed. For $A = 40-30$ kV/μs, the effective velocity at which the region of the impulse corona propagates is equal to $3-5 \times 10^6$ cm/s and the effective velocity of growing of a push-like leader is on the order of $1\cdot0 \times 10^6$ cm/s. The lengthening of the negative leader finally gives rise to breakdown which is typical of the standard wave. Thus, if $A = 40-3$ kV/μs, the development of breakdown has the following stages: (1) the initial impulse corona and the repeated impulse corona, (2) the pushlike leader, and (3) breakdown similar to that on the standard wave.

In this scheme, a rather intensive impulse corona, which increases the space charge near the channel tip, develops before the above-mentioned breakdown, and delays the development of the negative leader; V_d is higher than that for the diagram in *Figure 10B*. This accounts for the increase of the branch which is to the right of the minimum value V_d.

Summarizing the data on negative spark development, it is possible to distinguish the main elements which precede breakdown: (1) the impulse corona (initial and repeated), (2) the stepped leader of the spark, (3) the negative leader (continuous and push-like), (4) the space leader, (5) the positive leader, (6) the final jump, and (7) the main channel.

Some conclusions

(1) It has been shown that the change in the steepness of the voltage wavefront (A) in the −rod +plane gap causes the mechanism of the spark development to change essentially. The degradation of the negative leader, the tip of which circumscribes the solid line (*Figure 10A*) at $A < 40$ kV/μs into the push-like leader, is of particular interest. Its appearance resembles that of a scheme of a similar process of lightning mentioned by Walter (1903) and Toepler (1926). The theoretical interpretation of this phenomenon was given by Griscom (1958). This fact permits the conclusion that lightning develops on a rather slowly rising wave of voltage; this, in its turn, makes it evident that for calculation purposes, average gradients determined from the U-type voltage-time characteristic must be chosen, and not those for standard waves as it is accepted to do.

(2) The continuous negative leader develops in the −rod +plane gap on oblique waves at $A > 40$ kV/μs till the stage thread-like luminosity if $U_t - U_0(x) < 0$, where U_t is the instantaneous electrode potential and $U_0(x)$ is the discharge voltage when applied for some length of time ('prolonged action' of voltage) between the leader tip and the plane.

(3) As found by the authors, the continuous positive leader also develops in the +rod −plane gap on oblique waves at $A > 10$ kV/μs if $U_t - U_0(x) < 0$.

On the wavefronts close to standard, the positive leader will propagate in the +rod −plane gap if $U_t - U_0(x) > 0$ (Akopyan *et al.*, 1956, 1958).

(4) If $A < 100$ kV/μs, a phenomenon of great interest takes place: after the negative leader and the space leader join (*Figure 10B*), a jump-like lengthening of the former takes place, accompanied by a rapid transferring of the electrode potential to the negative leader tip, which results in the impulse corona. Simultaneously, the negative leader tip rapidly propagates toward the plane.

A similar process can be also seen in *Figure 10C* ($A < 40$ kV/μs). In this figure, the above-described phenomenon is observed to start at points P_1 and P_2 when the push-like negative leader develops.

Further, in both cases (*Figure 10B, C*), after the branches of the impulse corona or the stepped leader touch the plane, the positive leader develops from it. Then the leader tips approach one another, which results in the final jump and the main channel formation.

(5) Optical studies of the spark by the Boys camera and by other similar apparatus, which had been carried out until recently, were imperfect because of their limited speed of scanning and insufficient light sensitivity. In this connection, serious mistakes were made by a number of authors when describing the spark development; for instance, when describing the structure of the impulse corona and the velocity of its development, succession of the spark stages in time, especially in the −rod +plane gap, etc. The widespread extrapolations made from the mechanism of the spark development to the processes occurring in lightning and, expecially, in its leader stage, are not well justified. This accounts for the wide variation in its characteristics such as, the stepped leader channel diameter, the pilot-leader structure (whether it propagates uniformly or by jumps of the impulse corona) and others.

When analysing lightning, an optical study of it by the E.O.C. should probably be taken as a basis; it is hoped that a better correlation of the data on the two phenomena will be obtained if eocograms on the negative spark are available.

Summary
The paper deals with the description of the method for investigation of a long spark in rod – plane gaps with application to the rod of negative voltage waves of different front duration. Optical phenomena were recorded with an electron-optical converter; recording of electric processes was made by means of a multi-ray electron oscillograph.

The performed tests established a detailed picture of the negative spark development in gaps 1–3 m in length with voltage waves of different front steepness. New stages of discharge were discovered and sequence of spark development stages and quantitative characteristics of some discharge stages were determined. The mechanism of a long spark was found to depend essentially on the steepness of the wave of voltage applied to the gap. Of special interest is the transformation of the smoothly developing channel of the negative leader into a push-like leader that is again transformed into an uninterrupted process in the final portion of the discharge. Channels which arise in the gap and are then joined with the channel of an ordinary leader which starts from a negative electrode were registered. This phenomenon is of value when considering a possible mechanism of lightning.

The comparison of the mechanism of long spark development and that of lightning showed considerable difference between these processes. It is concluded that lightning mechanism should be experimentally studied in detail with devices of the kind used in our investigation.

References

Akopyan, A. A., Larionov, V. P. and Torosyan, A. S., *Electricity* **5**, 14, (1956); **6**, 33, (1958).

Bazelyan, A. M., Brago, E. N. and Stekolnikov, I. S., *Dokl. Akad. Nauk S.S.S.R.* **133**, 550, (1960).

Griscom, S. B., *Trans. A.I.E.E.* **7**, 919, (1958).

Park, J. H. and Cones, H. H., *J. Res. Natl. Bur. Std.* **56**, 201, (1956).

Schonland, B. F. J., *Proc. Roy. Soc.* (*London*) A, **164**, 132, (1938).

Schonland, B. F. J., *Proc. Roy. Soc.* (*London*) A, **220**, 25, (1953).

Schonland, B. F. J., Malan, D. I. and Collens, H., *Proc. Roy. Roy. Soc.* (*London*) A, **152**, 595, (1935).

Stekolnikov, I. S., *Izv. Akad. Nauk S.S.S.R., Otd. Tekhn. Nauk* **7**, 150, (1957).

Stekolnikov, I. S. and Shkilyov, A. V., *Dakl. Akad. Nauk S.S.S.R.* **136**, 803, (1960).

Stekolnikov, I. S. and Shkilyov, A. V., *Dakl. Akad. Nauk S.S.S.R.* **145**, 782, (1962)(a).

Stekolnikov, I. S. and Shkilyov, A. V., *Proc. Intern. Conf. Central Elec. Res. Lab., Leatherhead*, (1962)(b), p. 27.

Stekolnikov, I. S., Brago, E. N. and Bazelyan, A. M., *J. Teckn. Phys.* **32**, 993, (1962).

Toepler, M., *Mitt. Hermsdorf-Schomburg Isolatoren* **25**, (1926).

Wagner, C. F. and Hileman, A. R., *Power Apparatus and Systems* **52**, 682, (1961).

Walter, B., *Ann. Physik. Chem.* **68**, 776, (1899).

Walter, B., *Ann. Physik. Chem.* **10**, 393, (1903).

Introduction to Papers 21, 22 and 23 (Schonland, Malan and Collens, Extracts from 'Progressive lightning, Part 2'; Schonland, Extracts from 'Progressive lightning, Part 4—The discharge mechanism'; Schonland, 'The pilot streamer in lightning and the long spark')

The first two of the papers (21 and 22) from which extracts are given were part of a series on Progressive Lightning by Schonland and his associates which established the main visible features of a lightning flash. Only part of the paper describing the experimental work and part of the earlier theoretical paper is given here. The extracts selected describe in particular that part of the flash known as the 'stepped leader stroke'. In paper 22, the author found it necessary to postulate that the stepped leader stroke was preceded by a so-called 'pilot leader' or 'pilot streamer' which provided a volume of ionisation in front of an arrested stepped leader. Paper 23 gave a revised version of the theory, based on the new experimental data acquired between 1938 and 1953. (Schonland's schematic diagram of the development process is also shown in paper 20.) Other theories have been developed by Bruce (1944) and Komelkov (1947 and 1950), and in 1961 Wagner and Hileman formulated a theory which retained some of the main features of each of the earlier theories.*

Recently, Aleksandrov has reconsidered the mechanism of a spark discharge from a negatively charged point and of lightning. The central idea of his calculations is that the gas in an arrested leader step is heated by collisions between the gas molecules and energetic electrons, and a highly conducting channel is formed. This leads to an increase in the field between the end of the channel and earth and hence to the growth of a streamer from the tip of the channel. It is perhaps a little early to assess the long-term importance of Aleksandrov's work, but it is clear that more needs to be done to correlate studies of lightning with the results of laboratory experiments.

---

* A useful discussion of the various early theories of stepped leaders is given by Uman in the final chapter of his recent book *Lightning*. (McGraw-Hill, 1969). The chapter also includes brief discussions of the theories of dart-leaders and return strokes.

244 *Introduction (Papers 21–23)*

Further Reading

Malan, D. J., *Endeavour* **18**, 61, (1959).
Bruce, C. E., *Proc. Roy. Soc.* **A183**, 228, (1944).
Komelkov, V. S., *Bull. Acad. Sci. U.S.S.R.* **6**, 851, (1950).
Wagner, C. F. and Hileman, A. R., *A.I.E.E. Trans.* **80**III, 622, (1961).
Aleksandrov, G. N., *Soviet Physics-Tech. Phys.* **12**, 203, (1967).
Berger, K. and Vogelsanger, *Bull. Schweiz. Elect.* **57**, 1, (1966).
Uman, M. A., *Lightning*, McGraw-Hill, New York, (1969).

21

Extracts from *Progressive lightning—Part 2*

B. F. J. SCHONLAND, D. J. MALAN AND H. COLLENS*

1. Introduction

An account has previously been given of some results obtained with the Boys camera in the study of lightning discharges and of some deductions as to the nature of these discharges.† The present paper is concerned with further studies in the same field with improved technique and apparatus.

The material now available includes 41 photographs comprising 95 separate flashes and over 200 lightning strokes. These have been obtained from 13 different thunderstorms over a period of three years, and may therefore be considered to offer a fair sample of the lightning discharges of Southern Africa. We have derived from them a general picture of the nature of the discharge and the manner of its development which seems to be valid for the great majority of the cases studied.

This would suggest that the picture is applicable to lightning observations elsewhere, in temperate as well as tropical regions. It is therefore of interest to note that the results described in the previous paper have been confirmed by observations made in the Northern United States (lat. 42° N.) by K. B. McEachron and others‡ with a Boys camera. Further support for this view is to be obtained from the early observations of Walter, § whose work with a slow-moving single-lens camera is well known.

It is thus probable that the interpretation of our results which follows applies to flashes from storms in temperate as well as tropical climates. In temperate regions, however, a type of flash which is rare in South Africa and perhaps not so rare elsewhere—discharge from a positively charged cloud volume—may exhibit interesting and important differences in development.

The present paper is devoted to an analysis of the main facts which are likely to enter into any general discussion of the lightning discharge and its effects. The theoretical questions raised will shortly be dealt with in a further publication, as also will a few exceptional discharges presenting unusual features.

* Reproduced from *Proceedings of the Royal Society*, **A152**, 595, (1935) by permission of The Council of the Royal Society, London.

† Schonland, B. F. J. and Collens, H., *Proc. Roy. Soc.* **A143**, 654, (1934), referred to as part I.

‡ *Elect. (Cl.) J.* **31**, 251, (1934).

§ *Ann. Physik* **10**, 393, (1903), and *Jahrb. Hamburg wiss. Anstalten* **20**, (1903). We regret that these observations and deductions escaped our notice until recently. They are further discussed in sections 6 and 7.

It is necessary to point out that practically all the discharges which have been photographed occurred between the base of a thundercloud and the earth. Some information has been obtained about discharges which fail to reach the ground (section 10), but the most frequent type of discharge, that taking place entirely within the cloud, cannot be dealt with by these methods at all.

2. Apparatus and technique

For successful photography with the Boys camera, it is necessary to operate the device in the country, away from the lights of towns, for the camera must be exposed for some time before a photograph is obtained. It is therefore preferable to drive the rotating lenses by hand, and the difficulty of determining the speed of rotation is solved by operating just beyond one of the critical frequencies of vibration of the camera.

The type of camera used in most of the present work has already been described (part I, p. 654). The speed of rotation was about 3000 revs/min, nearly twice that previously employed. With good conditions this means a resolving power in time of $0.6 \mu s$ (part I, p. 656). For flashes 12 to 95, the photographs from the Boys and the fixed camera were supplemented by others from a slow-moving camera, working off the same drive as the fast Boys instrument. This camera consisted of a single lens, making one revolution for every 59 of the two-lens instrument. It was originally provided to give the order of appearance of the successive strokes of the discharge, but turned out extremely valuable in that it clearly showed the slower leaders to strokes.

A form of Boys camera in which the two lenses are fixed and the film revolves on the inside of a drum has recently been employed by us (flashes 82 to 87).* This arrangement gives resolving powers as low as $0.3 \mu s$, since the relative velocity of the two images of the discharge is 35 m/s. This form of Boys camera has several advantages, for it gives a linear record instead of a circular one, and the length of the linear track is 2·3 times as great. There is consequently less overlapping of the stroke images and less fogging from cloud discharges.

On many of the later photographs, *e.g.* flash 76, one lens has been open to $f/6.3$ and the other stopped down to $f/16$. This arrangement allows of photometric study of the relative intensity of the light from different parts of a stroke, the leader and the main return portion, for example.

In discussing the records we have, as before, allotted to each flash a number and distinguished the strokes forming one flash by letters. After flash 11, these letters correctly indicate the actual place of the stroke in the series; before flash 11, since no slow camera was available, they were allotted at random.

3. General account of the nature of flashes to ground

In this section an account is given of the general features of the lightning discharge which are common to most of the photographs studied, the evidence for the picture presented being discussed later.

Each lightning flash to ground consisted of a series of separate strokes ranging in number from 1 to 27. The time of separation was very variable, even in the same series, and values of from 0·0006 s to 0·53 s have been found.

---

* Constructed by Ross and Co. to the design of Sir Charles Boys, and slightly modified. This camera was very kindly presented to the South African Institute of Electrical Engineers by Mr. Bernard Price, O.B.E. For the fitting of the driving gear and auxiliary cameras we wish to thank Mr. C. J. Monk and Mr. F. C. A. Crewe.

The first stroke of each series differs from the others in two respects, which make it easy to identify. It is usually much more intense than the succeeding strokes and is always much more heavily branched.

Each stroke is composite, and consists of a leader portion travelling from cloud to ground and a main return portion moving faster and more brightly in the reverse direction. The leader to the first stroke consists of a series of streamers moving downwards in a step-by-step manner, and is hence referred to as '*stepped*'. The length of each step is about 50 m, and after completing a step the tip of the streamer appears to pause for a time of the order of 100 μs, whereupon the streamer extends still further, the new step being much brighter than the rest of the streamer.

The prolific branching of the main part of the first stroke of a series arises solely from downward branching in the stepped leader which precedes it. The course followed by the steps decides the path which is followed generally by all the strokes making up the flash and the zig-zag nature of the channel arises from changes in the direction adopted by successive steps.

The leaders to the *subsequent strokes* of the series are in general of the dart-like character previously described. They follow the path blazed by the first leader but usually do not branch. When branches are formed by either first or successive leaders they are subsequently more brilliantly illuminated during the return stroke.

Occasionally, however, when there has been an unusually long interval before the occurrence of a 'subsequent' stroke, the lower end of the usual dart leader becomes stepped like the leader to a first stroke. This type of leader is termed *dart-stepped*.

The second, upward-moving, return portion of each stroke follows directly upon the arrival of the leader part at the ground. In the course of its journey the return part branches outwards and downwards along the forks blazed by the leader. In some flashes, as at y in *Figure 1*, the leader is still moving out along a branch when the return part catches it up. In others, as at x in *Figure 1*, the leader branch has ceased to develop some time before the return portion starts. The very bright luminosity of the return stroke has its longest duration and greatest intensity at the ground and often shows a marked decrease in intensity at successive branches from the bottom upwards.

The main features of a straight vertical discharge as they would be shown on an endless film by one lens of a Boys camera with horizontal displacement are represented diagrammatically in *Figure 1*, which is not drawn to scale. The mean values of the various time intervals make it clear why several camera speeds are

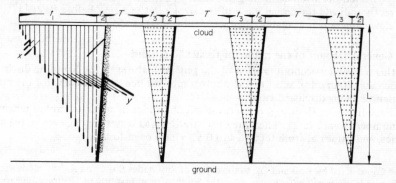

Figure 1. Average values: t_1 .. 0·01 sec; t_3 .. 0·001 sec; t_2 .. 0·00004 sec; T .. 0·03 sec; L .. 2·0 km.

necessary to obtain the fullest information about the discharge. The Boys method of studying the progress of a stroke depends, of course, upon the use of two records like *Figure 1* with lens displacements in opposite directions.

Figure 2.

The type of record given by the slow-moving camera for the first stroke of *Figure 1* is illustrated in *Figure 2*. Results of this kind were first obtained by Walter (*loc. cit.*) with a slow single-lens camera in 1902, and similar effects are to be found on photographs taken by Larsen in the same way. In both cases, however, the camera speeds were too slow to give the stepped-leader processes of *Figure 1*.

5. Leader strokes

(a) Occurrence of the leader-return stroke combination

In order to determine whether the leader-main stroke sequence was the most frequent, if not the invariable, mode of development in these discharges, the data were analysed as shown in *Table 3*, which indicates the numbers of first and of subsequent strokes showing leaders.

Of the 'good' pictures, 93% show leaders, while of 'good' and 'fair' combined the fraction is 73%. Including pictures classed as 'poor' we find 94 leaders to 158 strokes.

Where no leader portion of a stroke was observable, the subsequent main return portion always travelled upwards from ground to cloud as usual, and this fact, coupled with the figures in *Table 3*, leads us to conclude that the leader was too faint to be recorded. All the return portions of the strokes photographed by us have been found to travel upwards, and of the 33 complete flashes analysed in *Table 3*, 28 show leaders on one or more of their strokes.

In these thunderstorms, therefore, the leader-return stroke sequence was in the great majority of cases, if not always, the characteristic feature of discharges to ground.

(b) First stroke leaders

The leaders to first strokes can usually be most easily recognized on the slow-moving camera, where they have the dart-like appearance shown in *Figure 2*. On the fast Boys camera they are difficult to see unless the discharge is near, for the series of steps in their progress may occupy almost the full circumference of the circular track, much of which is taken up with other strokes and general fogging. On the best pictures we can see the whole course of these stepped leaders, but on many others only a portion is visible, depending on the amount of background present. Here recourse is had to the slow-moving camera to show that the leader track was actually present all the way down from cloud to ground.

Table 3—Occurrence of leaders

Photography	First Strokes		Subsequent Strokes		All Strokes	
	Number	Leaders	Number	Leaders	Number	Leaders
Good	15	15	26	23	41	38
Fair	18	9	41	26	59	35
Totals	33	24	67	49	100	73

Table 4 summarizes the manner in which information of this kind has been collected. In the first column we have described the photographs from the point of view of the possibility of detecting the rather faint steps on the fast camera record.

Table 4—Leaders to first strokes

Photography	Number of first strokes	Leaders observed on			
		Slow and fast cameras	Slow only	Fast only	None
Good	15	7	5	3	0
Fair	18	3	1	5	9
Poor	19	0	3	0	16

No case has been observed in which the leader to a first stroke has been other than stepped when examined with the fast camera.

(c) Subsequent stroke leaders

The leaders to subsequent strokes are generally of a continuous dart-like character, as described in part 1, p. 662, but in at least three discharges we have definite evidence of the 'dart-stepped' form, in which the dart changes at its lowest and slowest end to a stepped leader. The steps in such cases, as shown in *Table 5*, are very short, about 10 m long, and the pauses between them about 8 μs, so that it is not easy to detect them except under good conditions. It is possible that the effect occurs in other cases as well and that what has been described as the dart-leader process is actually only a special case of the stepped leader to the first stroke. In *Table 5* the length found to be stepped, shown in column 5, is the length nearest the ground.

Table 5—Dart-stepped leaders to subsequent strokes

Flash	Stroke	Velocity (cm/s 10^7)	Total length km	Length stepped km	Step length metres	Pause time μs
67	*b*	1·9	2·02	0·3	9·0	7·4
64	*b*	1·2	2·3	0·7	10·0	9·0
75	*d*	1·0	3·0	0·7	7·4	7·4

The detailed discussion of stepped leaders, in particular of those to first strokes, is resumed in section 8.

6. The velocities of leaders

(*a*) *Effective velocity of stepped leaders*

The stepped leaders to first strokes (sections 3 and 8) proceed in a series of jerks to ground, and consequently their *effective velocity* v is considerably less than the velocity V with which the streamer producing each fresh step advances. The effective velocity is obtained by dividing the total path length L by the time occupied over the path T and the values found for this quantity are given in the distribution *Table 6*. The table contains corresponding data for a number of stepped discharges from the cloud which do not reach the ground (*air-discharges*), and for the stepped portions of dart-stepped leaders to subsequent strokes.

The mean value of v for leaders to ground from first strokes is $3·8 \times 10^7$ cm/s, but about half the values listed lie within the range $1·0$ to $3·0 \times 10^7$ cm/s. The above are two-dimensional path velocities and should be increased by 30% to obtain the actual velocities in three dimensions (part I, p. 660). The values for the five air discharges and for dart-stepped leaders are higher than for ordinary stepped leaders.

The most probable length of time taken by a stepped leader to traverse 1 km of air would, from *Table 6*, be $0·007$ s, the range being from $0·01$ to $0·0008$ s. For three first-leader processes between a cloud at an undetermined height and the ground, Walter* gave durations of $0·02$, $0·009$, and $0·007$ s.

Table 6—Effective velocities of stepped leaders

Velocity range (cm/s x 10^7)		First stroke leaders to ground	Air discharges	Stepped parts of dart-stepped leaders
From	To			
1·0	2·0	9	0	0
2·0	3·0	3	0	0
3·0	5·0	3	0	0
5·0	7·0	3	1	0
7·0	9·0	3	3	0
9·0	11·0	1	0	1
11·0	13·0	1	1	0
13·0	15·0	0	0	1
18·7		0	0	1

* Walter, *loc. cit.*, pp. 21 and 30.

8. The nature of stepped leaders and air discharges

(a) *Some special cases*

In amplification of the descriptions given in section 3, three of the twenty-four cases of stepped leaders and air discharges, section 10, will be described in some detail.

In *Figure 10*, are reproduced the records of the four strokes of flash 66, which passed between cloud and ground at a distance of 12 km from the cameras. The leader process for 66*a* was so slow that it was only satisfactorily recorded on one of the Boys camera lenses, and this record is reproduced from an intensified

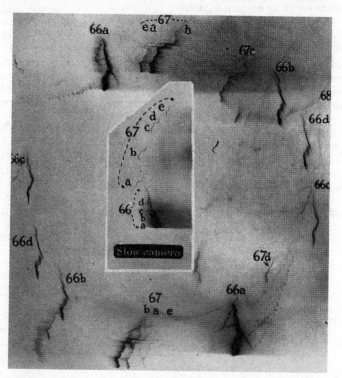

Figure 10.

positive copy in *Figure 11*, with a sketch diagram to scale in *Figure 12*. The clarity of the record of the main stroke has been unavoidably spoilt in intensification and the original picture is therefore shown in *Figure 11* as an inset.

In studying *Figures 11* and *12* it must be remembered that the lens motion was from right to left, and that the time is measured by displacement along the circular arc traversed by the ray through the centre of the lens, so that the illumination of a point at the top of the channel, for example, is to be found at various times along the arc *a*A. The leader began at *a* and ended below *e* at a time 0·0221 s later. The trunk of the leader channel followed the same course as the subsequent rapid main portion of the discharge (FDCBA) but in the opposite direction, and this course, owing to frequent pauses, was recorded by the moving lens along *abcde*. A branch developed from the leader at its starting-point *a* and extended in the same channel as that subsequently, brightly and rapidly, followed by the main stroke from A. Similarly at *b*, *c*, and *d*, the leader steps separate out to form the branches B, C, D. The last stages of the development of the leader trunk are visible at *e*, where it reaches to within 51 m of the ground in its penultimate step.

The leader reaches the ends of the upper branches A and B some 800 *μ*s before the main stroke arrives there, and so appears to cease to develop these branches some time before reaching the ground. Branches C and D, however, were still being pushed outwards by the leader at the time it reached the ground, and the leader steps for these two were caught up by the rapid upward-moving main stroke.

In this picture the streamers extending from the cloud-base downwards are just visible in certain of the stronger steps and then towards their lower ends only. Higher up they become wider and too faint to be distinguished from the general background.

Comparison of the two Boys camera records of the return main stroke shows that the total time taken by it to pass from F to A was 55·0 *μ*s.

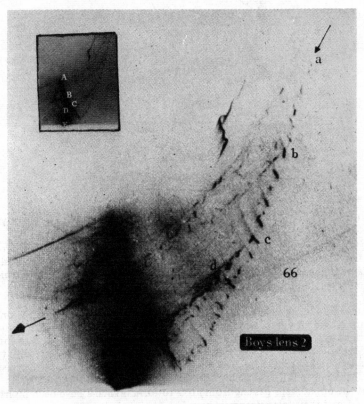

Figure 11.

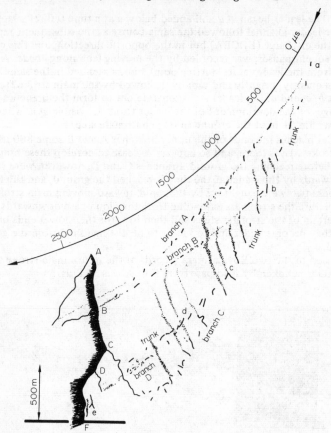

Figure 12.

Flash 76, *Figure 13*, and *Figure 14*, took place between cloud and ground at a distance of $2·0 ± 0·5$ km, so close that the upper part of the stroke was slightly out of focus for one of the Boys camera lenses and off the film for the other, *Figure 13(a)*. The flash, which consisted of a single stroke, was recorded on process film, the fine grain of which enabled a valuable record of the details of the leader to be obtained. These details are very clear in the lower half of the picture, which has been enlarged and intensified for *Figure 13 (b)* and is sketched to scale in *Figure 14*.

The leader trunk follows the course *abcdefg* corresponding to ABCDEFH on the rapid return stroke. The lens movement in *Figures 13* and *14* was along a circular arc from right to left. The leader branched at *j* to form branch J, and at *b* to form branch B, the subsequent steps of which form a series arranged along *bx*, with two side-forks developing from them. The details of these side-forks, one of which starts at *y*, have been omitted to avoid confusion. For the same reason only a certain number of the bright steps passing from *j* to *r* to form branch J have been inserted in *Figure 14*, and their faint streamers have been omitted altogether.

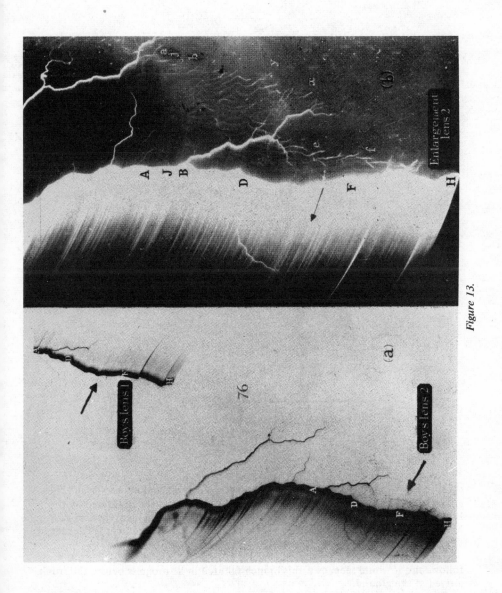

Figure 13.

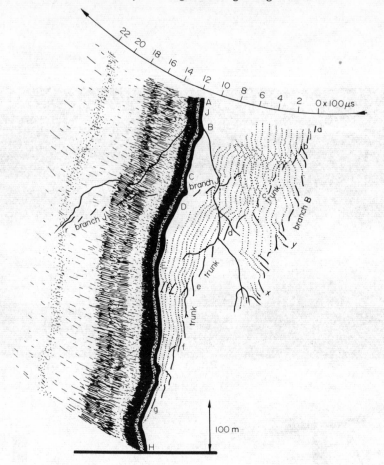

Figure 14.

The last step *r* of branch J traces out the final section of this branch and the rapid return stroke goes no farther than this step. The interval between *r* and the arrival of the return stroke is 18 µs, equal to the time interval determined by the Boys method between the start of the return stroke at the base and the end of branch J. Thus leader development of branch J ceased at the moment the leader trunk hit the ground.

The details of the leader steps at the end of branch B are more difficult to follow, and it would seem as if this branch faltered in its progress before the trunk arrived at the ground.

The streamers which end in these prominent bright tips are, as *Figure 13* shows, narrower and brighter at their lower ends, and they become difficult to follow as one passes upwards from their tips. This effect has not been shown in *Figure 14*. Normally, the luminosity of the bright tips appears to cease abruptly, but the penultimate step *g* in the leader trunk is exceptional in showing a faint continuation some 30 m long towards the ground. A similar effect, but to a much smaller extent, is shown by a few other steps higher up.

The lengths of the bright tips shown in *Figure 13* vary from 36 to 15 m, with an average value of 24·6 m. The variation in length must, of course, be due to some extent to the effect of motion in the line of sight. The time intervals between the ends of steps and the beginnings of their successors vary from 96 to 16 μs, with a mean of 52 μs. As the streamers approach the ground the step-lengths increase and the intervals decrease. Between *a* and *e* their average values are 21·5 m and 63 μs, and between *e* and *g* they are 27·0 m and 42 μs. The last interval of all, at *g*, is only 16 μs and the last step, as already mentioned, is unusually long. The effective velocity of the leader process is $3·4 \times 10^7$ cm/s from *a* to *e*, and $6·4 \times 10^7$ cm/s from *e* to *g*.

Other details of these steps are included in the next two sections.

(b) *Summary of details of leader steps*

The following summary contains a list of the main features of the leader steps, quantitative study of some of them being left to the next section. The statements made are based on the full material available to us,* and reference is made where necessary to *Figures 12* and *14* as illustrations.

(1) Each bright step appears as the termination of a fainter streamer extending the whole way down from the starting-point of the discharge, such streamers increasing in brightness and decreasing in width as their ends are approached.

(2) A bright step generally starts a little way back on part of the track formed by the previous step, so that the portion of the step which is entirely new is about 90% of the whole, *Figure 14*, step after *e*.

(3) The form of the channel blazed by a new step is not always followed in detail by the streamers of subsequent steps, the tendency being for the smaller twists and kinks to be smoothed out. In a few cases the form of the subsequent streamer is changed so completely as to suggest that a fork previously existed, the fainter branch of which was adopted as the channel in later stages, *Figure 14*, second step after *e*.

(4) A beaded effect at the beginning and end of the bright steps is frequently found, which in view of its position cannot be due to motion in the line of sight.

(5) Branching occurs as a division of one step into a fork, *Figure 12*, *c*, and *Figure 14*, *b* and *y*, the subsequent development of a branch taking place in steps simultaneous with those of the trunk. Short branches and branches near the top of the channel, however, may cease to develop some time before the main trunk steps have reached the ground. The start of a branch never occurs elsewhere than at the beginning of a bright step.

(6) Each fresh step takes in general a different direction to its predecessor, and it is unusual to find a sharp bend in the course of a single step. The tortuous form of a lightning discharge channel thus originates in the different directions taken by successive steps.

(7) The streamers and their tips are generally much brighter in their later stages, when they are approaching the ground, than earlier. It is not always possible to detect even the tips in the upper portion of the leader. This increase in brightness is associated with a considerable increase in effective velocity near to the ground.

(8) The brightness of the streamers and their tips is found to increase with increase in the effective velocity, v (section 6a), of the leader process. The three cases discussed at length in section 7a are discharges in which v is higher than the average.

* 24 cases in all.

(9) Some hesitancy is occasionally shown in the development of both branch and trunk steps. Thus the branch which started at *e* in *Figure 14* missed three steps and then re-blazed the first branch step twice. At the same spot the trunk development missed a step and made up for it by an exceptionally long jump the following step. Similar cases occur at *y* and at *f* in *Figure 14*, and at the step two below *c* on branch C of *Figure 12*.

(c) Velocity of leader streamers

To determine the velocity of the streamers which make up the leader steps one can either use the Boys method of comparing one rotating lens record with the other, or compare the track of the streamer, as recorded by one lens, with that of the return stroke, whose distortion is easily found by the Boys method. Difficulties, however, arise from the limited lengths of streamer tracks available for measurement and the lack of sharpness in their leading edges. Such comparisons have not so far yielded any evidence of distortion of the streamer track as a result of lens motion, and it is concluded from several measurements that the streamer velocity exceeds 5×10^9 cm/s.

Definite evidence that the streamers travel downwards is, however, afforded by the broadening of the upper part of their tracks. If we take the duration of luminosity, t, at a point distant d from the tip as being equal to the time taken for the streamer to travel this distance we again obtain, as an upper limit for the velocity, values of d/t of the order of 5×10^9 cm/s. Unfortunately, the method is too rough to indicate whether the streamer velocity varies at different points along the streamer or in different streamers.

The tips of the streamers, which from their sudden cessation of luminosity might be expected to show a decrease in velocity, exhibit no distortion due to lens motion and have velocities which are certainly greater than 10^9 cm/s.

(d) Relations between step-length, step-interval, and effective velocity

It is convenient to denote by $\bar{l}$ the average length, along the two-dimensional record, of the leader steps, by $\bar{t}$ the average time interval between steps, so that the quantity $\bar{l}/\bar{t}$ is an average velocity. If the time occupied in tracing the step-length $\bar{l}$ is small compared with $\bar{t}$, this velocity should be the same as the effective velocity v (L/T) with which the stepped leader advances. Comparison of v with $\bar{l}/\bar{t}$ shows them to be identical within the accuracy of measurement, as indeed would follow from the high value of the streamer velocity along $\bar{l}$ found in the preceding section.

Values of $\bar{l}$ and $\bar{t}$ have been obtained with sufficient accuracy from 11 first-stroke leaders, three cloud-air discharges, and three dart-stepped leaders. The remaining seven cases of the stepped process are too poor to give reliable data. The results are plotted in *Figure 15*, which shows the dart-stepped cases with crosses and the remainder with circles. The black circles indicate cases where the photography was good and the measurements could be made with more accuracy than for the open circle cases. It will be seen that while $\bar{l}$ varies from 10 to 206 m a ratio of 1/20, $\bar{t}$ for the first strokes only alters from 31 to 91 μs, a ratio of 1/3. The larger step-lengths are associated with the longer time intervals. The same type of variation in the course of a single leader has already been noted for flash 76.

To a rough first approximation $\bar{t}$ might be considered as constant, and hence it is that when in *Figure 16* the values of $\bar{l}$ are plotted against those of v, or $\bar{l}/\bar{t}$, the points lie fairly well on a straight line through the origin. A better fit, however, is obtained in *Figure 16* with the dotted line which cuts the horizontal axis at $v = 1{\cdot}0 \times 10^7$ cm/s, a value for which, as will be shown later, there is theoretical justification. From this line it follows that the relation between $\bar{l}$ and $\bar{t}$ is

$t = 80\,\bar{l}/(\bar{l}+800)$, where the units are microseconds and centimetres. The curve of *Figure 15* has been drawn from this equation.

The corresponding relation between $\bar{t}$ and v is $\bar{t} = 80\,(v - 10^7)/v$, where the units are microseconds and centimetres per second.

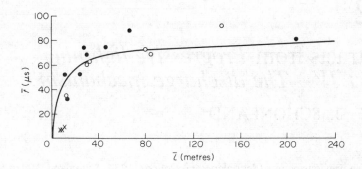

Figure 15.

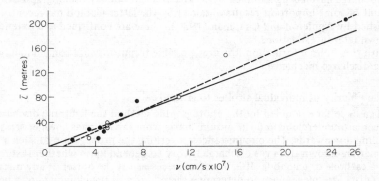

Figure 16.

22

Extracts from *Progressive lightning Part IV—The discharge mechanism*

B. F. J. SCHONLAND*

In three previous papers (Schonland and Collens, 1934; Schonland, Malan and Collens, 1935; Malan and Collens, 1937) in this series, the writer and his collaborators have described the results of a study of the lightning discharge to ground by means of the Boys and other cameras. During the last two years these studies have been continued with improved instruments, and the electrical changes taking place during the discharge have been examined by means of a cathode-ray oscillograph. An account of some important results obtained by the latter method has recently been published by Appleton and Chapman (1937). These are confirmed and extended by our own later investigations.

With this fuller information it is now possible to put forward some deductions as to the discharge mechanism.

1. The polarity of individual strokes to ground

In all cases so far examined by the photographic method the lightning discharge between a thundercloud and the ground starts from the cloud as a leader streamer travelling downwards. The circumstances governing the propagation of such a streamer are known from gas-discharge theory to depend upon whether it starts from a cathode or an anode. It is therefore necessary at the outset of any discussion of the discharge mechanism to determine the polarity of the cloud end of a lightning stroke.

As a result of numerous observations in many parts of the world it is now generally accepted that the total charge conveyed to ground in the great majority of discharges is negative. This would seem to indicate that the polarity of the discharge is in general such as to make the cloud end a cathode. It has, however, never been established that this conclusion holds in detail for each of the component strokes which make up a discharge. Indeed Norinder (1936) has reported that observations made with a cathode-ray oscillograph in Sweden show that a positive discharge stroke is sometimes followed by a number of negative ones.

For the South African lightning discharges which have been studied photographically, however, the conclusion suggested by previous studies of the nett change of field (Schonland, 1928; Halliday, 1932), that the discharge proceeds from a cloud cathode, has now been established for each separate stroke of a composite

* Reproduced from *Proceedings of the Royal Society*, **A164**, 132, (1938) by permission of The Council of the Royal Society, London.

flash to ground. Seventy such flashes comprising some 300 separate strokes to ground have been examined with a cathode-ray oscillograph whose resolving power extends to 50 μs. All the strokes of each flash were recorded photographically and in each case the field changes pointed unambiguously to the slow lowering and sudden destruction of a negative cloud charge.

2. General survey of the discharge process

Each of the successive strokes or partial discharges which make up a complete lightning discharge to ground takes place in two stages, the downward-moving leader stage being followed by an upward-moving return stage. These processes will be described as the leader and the return streamer, since they have the same properties as electrical streamers produced in the laboratory. Such a streamer is a conducting filament of ionized gas which extends its length by virtue of ionizing processes occurring in the strong field in front of its tip. It is electrically charged throughout its length* but is not at the same potential as the electrode from which it started, for there is a drop of potential along it. This drop of potential provides a field which drives a current through the stem of the streamer and this current serves to charge up newly formed sections of the stem to the potential necessary for further progress. The current continues at the tip of the streamer as a convection current due to the charge situated there and beyond the tip as a displacement current (section 7). Light is emitted by the streamer as a result of excitation processes at the tip (section 8). Apart from this the luminosity associated with the streamer is small and it can therefore be inferred that the field in the stem behind the tip is insufficient to cause excitation by electron impact.

It follows from section 1 that all leader streamers observed by us are negative or cathode streamers, and all return streamers are positive or anode streamers. The leader process in the case of strokes subsequent to the first is a continuous one and will be termed the dart streamer. That for the first stroke proceeds in a series of steps, each step requiring the development of a new step streamer.

The leader streamer of the first stroke lowers negative charge into the air and distributes it* over the conducting system formed by the leader channel and its branches in the manner shown in *Figure 1(a)*. It was suggested from the photographic studies (Schonland, Malan and Collens, 1935, p. 622) that this charge represented a considerable fraction of that tapped by the leader. This is confirmed by observations of the electrical field change, for Appleton and Chapman (1937) report that the leader or '*a*' stage observed by them causes a considerable change of the thundercloud moment. Our own oscillographic records show that the first leader lowers into the air about 85% of the total charge tapped by it.

Our observations also show that in the case of strokes after the first the fraction of the charge lowered is much less than in the first leader. This is to be explained by the fact that these leaders do not usually have branches. The leader system has therefore a smaller capacity and is less effective in lowering charge from the cloud.

The second or return stage of the discharge is initiated just before the arrival of one of the leader branches at the ground, by the upward passage of positive streamers from the earth. These streamers have been photographed (McEachron and McMorris, 1937), and the frequent presence of more than one can be inferred from the root branching shown at the base of many lightning channels (Schonland, Malan and Collens, 1935, p. 602). This stage is shown in *Figure 1(b)*.

* These statements, like others in this section, follow what appears to be the accepted and correct view as to streamer development, which it is intended to discuss more fully elsewhere.

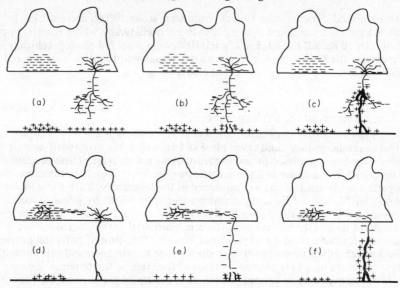

Figure 1.

Once contact is established between leader and earth, the positive return streamer passes rapidly up the leader channel. Such a streamer, as was first shown by Simpson (1926), must advance by drawing electrons, produced as a result of collision processes in front of its tip, inwards to the tip and passing them down to ground via its conducting stem. The relatively immobile positive ions left in front of the tip are then responsible for its rapid extension. This stage is shown in *Figure 1(c)*. The change of electric moment in this stage of the discharge is large and rapid. Its identification by Appleton and Chapman with the second '*b*', stage of the field change record is supported by our own observations.

The last portion of the return stroke process involves the removal and passage down the channel of the residual charge on the cloud centre tapped by the stroke. This stage, which is shown photographically by a continuance of channel luminosity after the return streamer has entered the cloud and which is of comparatively long duration, may be identified with part of the final or '*c*' portion of the field-change record.

It is implied in this description of the discharge process that separate strokes tap different centres of charge within the same thundercloud, a suggestion based upon evidence discussed in section 10. The general nature of the discharge process for a second stroke from a new charge centre is illustrated in *Figures 1(d)*, (*e*) and (*f*).

3. The leader process

The difference in the behaviour of leaders to first and to subsequent strokes implies a vital difference in the mechanism of their advance. The dart streamer invariably follows the path traced out by a previous stroke, even to shifting its track laterally if a wind has blown this path aside (Schonland, 1937). This and other features connected with its velocity under different conditions (section 4) indicate that its mechanism is that of streamer advance along a previously ionized channel, which is further discussed in section 4.

The stepped leader, on the other hand, is characteristic of advance into what is apparently virgin, unionized air and involves a different mechanism. Associated with it are two velocities, for while the step streamers advance at more than 10^9 cm/s, the effective velocity of the process as a whole is much lower, most frequently only $1\cdot5 \times 10^7$ cm/s. It has never been observed to fall below $1\cdot0 \times 10^7$ cm/s, though much lower values would not have escaped observation. This minimum velocity in the step process is an important clue to the mechanism involved in it, for it can be shown that there is strong presumptive evidence that it is the real velocity of a preliminary streamer which precedes the step streamer.

A negative streamer can be imagined to advance into virgin air in one of two ways, as a result of ionization produced either by electrons situated in its tip or by photoelectrons generated in front of its tip. In either case a lower limit to streamer velocity is set by the existence of a critical field strength, E_c, in front of the tip, below which ionization by electron impact cannot occur. The corresponding critical electron drift velocity, v_c must be the minimum velocity of advance of a negative streamer, since the streamer as a whole cannot move more slowly than the ionizing agents which produce its extension (Schonland, 1935; Goodlet, 1937).

It is usual to write the relation between v_c and E_c in the form

$$v_c = \sqrt{(2E_c e\lambda/\pi m)}$$

where, following Townsend, the assumption is made that every collision between electron and gas molecule is inelastic. Since $E_c \times \lambda$ is known to be sensibly independent of pressure between 54 cm Hg (the mean pressure involved in these discharges) and 76 cm, we substitute $E_c = 30\,000$ V/cm and $\lambda = 3\cdot8 \times 10^{-5}$ cm, the gas-kinetic electron mean free path at $20°$C and 76 cm Hg, and find for the critical velocity, v_c, the value $3\cdot6 \times 10^7$ cm/s.

Two factors combine to make this result too high. In the first place it is known that for nitrogen the value chosen for λ is too large by a factor ranging from $0\cdot8$ to $0\cdot3$ according to the velocity with which the electrons are moving (Klemperer, 1933, p. 290). Secondly, the assumption that all collisions are inelastic when the limiting velocity is approached cannot be correct. For purely elastic collisions the value of v_c is found to be less than that given above by the factor $\sqrt[4]{(\pi^2 m/4M)}$ (Compton and Langmuir, 1930, p. 221) or $0\cdot08$, m being the mass of the electron and M that of an 'air' molecule. Applying the first correction, v_c cannot exceed $1\cdot8 \times 10^7$ cm/s, while from the second argument it cannot be less than $1\cdot4 \times 10^6$ cm/s. It is difficult to see how any closer approximation to v_c can be obtained.

The observed minimum effective velocity of the stepped leader process, $1\cdot0 \times 10^7$ cm/s, will here be identified with v_c, the minimum velocity of advance of an actual negative streamer. This identification is supported by the observations made by Allibone and the writer upon negative leaders in the laboratory spark discharge which appear, as Goodlet has recently pointed out (1937), to show a similar minimum velocity of the order of $1\cdot5 \times 10^7$ cm/s.*

The present suggestion is equivalent to the statement that a true negative virgin air streamer travels continuously downwards in front of the step streamer processes with a velocity equal to the effective velocity of these. Upon this, so far unrecorded, streamer the steps are periodically superimposed. It follows that the step streamers, like the dart streamers of subsequent strokes, travel along a previously ionized channel provided by this slower *pilot streamer* which precedes them.

* The existence of a minimum velocity of this order for negative streamers serves as an additional criterion of polarity. The minimum for positive streamers, if it exists, is much lower, for the writer has recently photographed such streamers with a velocity of $1\cdot6 \times 10^6$ cm/s.

The suggestion finds immediate support in the relation which has been found to exist between the time of pause of a step streamer, t, and the length of the step, l, which is executed after this pause (Schonland, Malan and Collens, 1935, p. 616). For each individual case the ratio l/t is equal to the effective velocity at this stage of the process, that is to the velocity of the suggested pilot streamer. If t is long the subsequent step l is long and vice versa. The steps thus retrace and brightly illuminate an ionized channel formed during the pause period and cease when they have caught up with the tip of the pilot streamer. During the following pause, the origin of which is discussed in section 6, the pilot streamer once more forges ahead. *Figure 2* illustrates what would be observed on a camera with a fixed lens and a film moving from right to left if the pilot streamer, shown as a dotted line, could be recorded as well as the step streamers which follow it. An explanation of the small luminosity associated with this pilot and the consequent difficulty of recording it photographically or by the field-change method is given section 8.

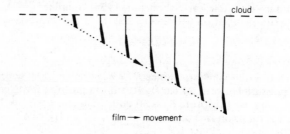

Figure 2.

It is possible in this way to offer a simple explanation of the tortuous nature and branching of the first leader channel. In the majority of cases the effective or pilot velocity of the process is less than $1 \cdot 5 \times 10^7$ cm/s, not very far from the critical minimum of 10^7 cm/s below which it cannot progress at all. Its direction will therefore be controlled by the structure of the electric field in the neighbourhood of the pilot streamer tip and by variations in the density of local space charge.

It may be noted that a stepped method of development in the case of Lichtenberg figures from a negative point has been inferred by v. Hippel (1933, 1934) under certain experimental conditions and has been interpreted by him in terms of a pause during which a pilot streamer (*vorentladung*) moves ahead of the main spark channel. A similar mechanism was proposed by Toepler (1926) to account for beaded lightning (*perlschnurblitz*).

So far the discussion has been limited to the case of pilot streamers travelling with the minimum observed velocity. Before it is possible to refer to the behaviour of such streamers at higher velocities it is necessary to consider the effect which photoionization will have upon its development in fields greater than the critical value E_c. This question is also involved in the mechanism of all streamers travelling along a previously ionized path. In the case of the lightning discharge there are three such types of streamer—the dart leader streamer to subsequent strokes, the return streamer to all strokes and, as has been suggested above, the step streamer which follows the pilot streamer in first leaders.

References

Appleton, E. V. and Chapman, F., *Proc. Roy. Soc.* A, **158**, 1, (1937).

Compton, K. and Langmuir, I., *Rev. Mod. Phys.* **2**, (1930).

Goodlet, B. L., *J. Instn elect. Engrs.* (1937).

Halliday, E. C., *Proc. Roy. Soc.* A, **138**, 205, (1932).

Hippel, A. v., *Z. Phys.* **80**, 19, (1933).

Hippel, A., *Naturwissenschaften* **22**, 47, (1934).

Klemperer, O., *Electronik*, p. 290. Berlin: Springer, (1933).

McEachron, K. B. and McMorris, W., *Gen. Elect. Rev.* **39**, 487, (1937).

Malan, D. J. and Collens, H., *Proc. Roy. Soc.* A, **162**, 175, (1937).

Norinder, H., *J. Franklin Inst.* **225**, 69, (1936).

Schonland, B. F. J., *Proc. Roy. Soc.* A, **118**, 233, (1928).

Schonland, B. F. J., *Nature, Lond.* **136**, 1039, (1935).

Schonland, B. F. J., *Phil. Mag.* **23**, 503, (1937).

Schonland, B. F. J. and Collens, H., *Proc. Roy. Soc.* A, **143**, 654, (1934).

Schonland, B. F. J., Malan, D. J. and Collens, H., *Proc. Roy. Soc.* A, **152**, 595, (1935).

Simpson, G. C., *Proc. Roy. Soc.* A, **111**, 56, (1926).

Toepler, M., *Mitt. Hermsd.-Schomb. Isolat. G.m.b.H.* **25**, 743, (1926).

23

The pilot streamer in lightning and the long spark

B. F. J. SCHONLAND*

1. General information on the pilot streamer

When the first leader streamer moves down from a thundercloud to earth, the breakdown of the air in front of its tip is considered to take place in two stages (Schonland, Malan and Collens, 1935; Schonland, 1938). In the first stage, a weakly ionized pilot streamer advances with a velocity of from $1·5 \times 10^7$ to $8·0 \times 10^7$ cm/s over a distance of from 10 to 80 m. The evidence for this streamer (Schonland, 1938) is indirect, for the streamer has not yet been recorded photographically. In the second stage the highly ionized leader, which has been stationary during the advance of the pilot, suddenly moves forward and overtakes it in a rapid bright step. The velocity of the step exceeds 5×10^9 cm/s, so that its duration is of the order of a microsecond. On reaching the end of the pilot streamer, the leader stops and the pilot moves forward again for a time which usually ranges from 30 to 90 μs before it is again overtaken by the leader.

That the same two stages, pilot followed by leader, occur in the breakdown development of a very long spark from a negative point was first shown by Allibone (1948). The pilot stage was found by him to be a corona streamer of large radius containing many fine filaments, so faint that it was best recorded by a Lichtenburg-figure technique. This streamer extended across the whole of the gap and into it developed subsequent narrow leader streamers from both electrodes. When these electrodes were rods, the ratio of gap-width to corona-streamer radius lay between 12 and 20. If the anode was an earthed plane, the divergence of the field caused the ratio to fall at the anode end to about 3.

Direct photography of this pilot streamer in the case of an impulsive 3 MV discharge has been achieved by Hagenguth, Rohlfs and Degnan (1951) using a quartz lens. The gap width was 5 m between rod electrodes and the radius of the pilot streamer 31 cm, a ratio of 16/1. From the records of current in the earth-lead and of potential variations at the cathode it can be estimated that the velocity of the

* Reproduced from *Proceedings of the Royal Society,* **A220**, 25, (1953) by permission of The Council of the Royal Society, London.

pilot streamer was 3×10^7 cm/s. The average field strength across the gap before discharge and hence the average gradient at the moment the pilot streamer crossed the gap was 6×10^3 V/cm.

Gaunt and Craggs (1951) have reported similar results for the long spark from a positive point at some 37 kV with reference to an earthed plate. The average gradient in air was $6 \cdot 0 \times 10^3$ V/cm. In nitrogen the ratio of gap width to radius was approximately 6/1. In air the radius appears to contract, but mention is made of a 'continuous luminous background'. Gaunt and Craggs found from direct measurement with photo-multipliers that the luminosity travelled across the gap with a velocity of approximately $1 \cdot 0 \times 10^7$ cm/s. From general indirect evidence they concluded that the current density in this streamer was low compared with that in the narrow streamers of electron-avalanche type.

The purpose of this paper is to establish quantitatively the identity of the pilot processes in lightning and in the long spark and to make certain deductions from this identity. From these a new explanation of the stepped process is derived.

2. The continuity of the current in leader and pilot streamers

Since the publication of the first studies, photographic and electrical, of the stepped leader a further and important piece of information about it has become available. This is the observation of Chapman (1939), confirmed by Malan and Schonland (1947), that there is no observable stepped electrostatic field change during the downward course of a nearby stepped leader. Dr D. J. Malan and Mr N. D. Clarence have recently made a special investigation of this point at my request and find that the step electrostatic field changes are of magnitude less than 1/10 of the continuous field changes in the intervals between the steps.

It follows that the fast bright step does not carry with it in one large pulsation the charge needed to create the next section of the leader, but that the transport of charge is a continuous process, proceeding without important interruption during the whole time of advance of the pilot streamer. The pilot streamer must therefore carry an excess negative charge in its stem.

That this is a reasonable deduction can be seen by considering the current which the bright step would have to carry if the pilot were not charged. The charge Q' on a step of 20 m in length is $20/(5 \times 10^3)$ of the total charge on a leader 5 km long. Q, as will be shown in the next section, is at least 4 C, hence $Q' \geqslant 0 \cdot 016$ C. For this to be carried forward in less than 1 μs would require a current exceeding 16 000 A. The luminosity of the step process is far too weak to make it likely that the step carries a current of this magnitude.

The absence of any appreciable electrostatic field change due to the rapid steps of the leader is not inconsistent with the considerable radiation field produced by them, for the radiation field depends upon the rate of change of the current and not on the current itself.

Records of the waveform of atmospherics enable a comparison to be made of the average amplitude of the step-radiation pulses from a leader and of the radiation pulse due to the start of the return streamer. Professor D. B. Hodges, who has recently made the necessary measurements in this Institute, finds that the ratio of these two amplitudes lies between 1/20 and 1/10. Since this ratio is the same as that of *di/dt* for the average step and for the return streamer, and since the latter is known from records of the time variation of the current in direct strokes to towers to have a most frequent value of 10 kA/μs (Hagenguth and Anderson, 1952, table III), *di/dt* for the average step must be less than 1 kA/μs. A step of duration 1 μs thus involves a current change of less than 1000 A, which would be a reasonable value to account for the luminosity observed. It would involve the transport of less than 0·001 C in the step process, which is less than 1/16 of that already carried forward by the pilot. Thus the electrostatic field change due to the step would be difficult to detect, though its radiation field change would be quite significant.

The results reported above would seem to invalidate all 'sudden pulsation' explanations of the stepped leader such as those of Szpor (1942) and Komelkov (1950), both of which treat the leader channel as a relaxation oscillator, and that of Bruce (1944), which requires a glow-arc transition in the pilot channel. The results also offer further evidence of the actual existence of the pilot streamer.

3. The charge per unit length of leader and pilot and the current carried by them

Before the step process can be examined quantitatively, an estimate must be made of the charge q carried on unit length of both pilot and leader. It has been shown by Schonland, Hodges and Collens (1938) that the total charge Q on the first leader when it has extended its full length L to the ground is approximately 60% of the charge originally present on the section of the cloud which is tapped by it. From the measurements of Wilson (1921) and others, the total charge disposed of in one flash is known with some certainty to have a modal value of 20 C. Since the most frequent number of separate strokes is three, $Q \approx$ 20 x 0·6/3 or 4 C.

The average height of the region in the cloud which is tapped by these first leaders is 3·5 km above the ground (Malan and Schonland, 1951). Allowing 30% for the tortuous nature of the channel, the modal value of L may be taken as 5 km.

The quantity q may therefore be estimated to be 8 x 10^{-6} C/cm or 2·4 x 10^4 e.s.u./cm.

The modal value of the time T taken by a first leader to reach the ground is known from many records of leader field changes obtained by Dr D. J. Malan and the writer in Johannesburg to be 0·0125 s. Hence the leader current $I = Q/T$ has a most frequent value of 320 A. This is in satisfactory agreement with direct observations by McEachron (1939) of the current in upward-moving stepped leaders, both positive and negative, which ranges from 50 to 650 A.

The most frequent value of the pilot velocity is 4 x 10^7 cm/s. The step length corresponding to this velocity is about 20 m (Schonland *et al.*, 1935).

4. The radius of the leader and pilot streamers

Three separate general arguments show that the lightning pilot and the leader behind it must have a radius of several metres.

(*a*) Since the current I, apart from a temporary burst during the step itself, is continuous from leader to pilot, it can be written $I = qv = \frac{1}{2}Xvr$, when v is the velocity of the pilot, X the field in front of its tip and r its radius. The first of these expressions for I states that each centimetre of the developing pilot requires the supply of a negative charge q. The second expression is that given by Rudenberg (1930) for the displacement current in front of the moving streamer tip, corrected for an error in the numerical coefficient. Thus

$$r = 2q/X \tag{1}$$

For a given value of q, the maximum possible value of r is reached when $X = 3 \times 10^4$ V/cm, which is well known to be the minimum field for streamer advance. Substituting $2\cdot4 \times 10^4$ e.s.u./cm for q and 100 e.s.u. for X in equation (1), $r_{max} = 4\cdot8$ m. In this minimum field, v would be about $1\cdot0 \times 10^7$ cm/s (Schonland, 1938).

The most frequently observed value of v for the pilot process is, however, $4\cdot0 \times 10^7$ cm/s, and to find the corresponding value of r requires consideration of the manner in which v varies with X. If the advance of the pilot streamer is due solely to direct collisional ionization by electrons drifting in the field X with velocity v', well-known expressions for v' show that when it is 4×10^7 cm/s, X must be about 2×10^5 V/cm with an uncertainty factor of 2 or 3 (Loeb and Meek, 1941, p. 44). There is, however, much evidence that photons emitted by excited molecules create ionization in front of the drifting electrons, and hence that the velocity v for a given X is greater than v' (Cravath and Loeb, 1935; Schonland, 1938). Thus X must be less than the value required to make $v = v'$, though exactly how much less is uncertain. To get an estimate of r, we will adopt the factor 3 given by Loeb and Meek and thus substitute $X = 200$ e.s.u. (6×10^4 V/cm) in equation (1). The radius of the pilot streamer is then found to be $2\cdot4$ m.

(*b*) Although the channel of the lightning leader behind the step portion carries a large current, its luminosity is found to be very weak. The channel is observed to retain its conductivity, without rejuvenation by new ionizing processes, for as long as $0\cdot0125$ s, the average time the leader takes to reach the ground. These facts indicate that the leader channel is thermally ionized and has arc characteristics. It must be regarded as a good conductor, and its channel radius must be such that the field at its surface is everywhere less than that which would give rise to corona, for otherwise it would enlarge its radius still further.

This argument, which was first put forward by Goodlet (1937), requires that $2q/r \approx 100$ e.s.u., from which the radius of the leader channel (for $q = 2\cdot4 \times 10^4$ e.s.u./cm) is found to be $4\cdot8$ m, of the same order as that derived in (*a*) for the pilot channel.

This conclusion is strongly supported by observations of the spark-leader channel made by Komelkov (1947), who has shown that it exhibits two sharply defined zones: an internal leader core of relatively small diameter ($0\cdot2$ cm) but intense luminosity, and an outer 'streamer' or 'corona' zone of radius much greater than that of the leader channel, but much weaker in luminosity. He found the longitudinal field in the leader core to be about 55 V/cm and showed that the radial field at the surface of the outer corona zone was 3×10^4 V/cm.

(*c*) The observations that the conducting leader streamer periodically remains stationary for a time of the order of 50 μs can be interpreted *a priori* in three ways:

(i) in terms of a periodic failure of the cloud charge to supply current to the leader.

(ii) in terms of a periodic loss of conductivity by the leader channel behind its tip, and

(iii) in terms of a periodic decrease of the field in front of the leader tip below the value at which it can advance at all.

On further examination the first of these explanations is rendered unlikely by the regularity of the phenomenon and is difficult to accept now that it is known that the region of the cloud tapped by the leader has been converted into a good conductor by the *j* process which precedes the emergence of the leader (Malan and Schonland, 1951). Furthermore, although this *j* process is repeated in successive leaders, the step phenomenon is observed for the first leader only.

The second explanation is ruled out by the conclusion in (*b*) above that the channel is thermally ionized. Thus the only explanation which remains is that the field in front of the leader tip is periodically insufficient to enable it to advance *in the form of a thermally ionized channel*, though it is always great enough to permit the development of the pilot streamer.

The lowest value of the electric field which will permit the advance of any streamer at all is 3×10^4 V/cm. For the advance of a thermally ionized streamer the field must certainly be greater than this but not necessarily very much greater, since the probabilities of collisional ionization and of excitation increase very rapidly with field strengths greater than 3×10^4 V/cm. For the leader itself to be arrested it is therefore reasonable to suppose that the field strength X in front of its tip must be less than 6×10^4 V/cm. Since the starting out of the pilot requires $X \geqslant 3 \times 10^4$ V/cm, X must lie between these two limits. From this conclusion the value of r may be estimated. X is mainly due to the charge on the hemispherical tip of the leader; the contribution from the farther charge on the channel behind the tip can be ignored for approximate calculation. As q is the charge per unit length, the charge on the tip is qr and the charge per unit area on it is $q/2\pi r$. Thus $X \approx 2q/r$ and $r \approx 2q/X$. Since $q = 2\cdot4 \times 10^4$ e.s.u. and X is between 100 and 200 e.s.u., r must lie between 2·4 and 4·8 m.

These three arguments indicate that r for both pilot and leader is not far from 2·5 m, and this value will be adopted in subsequent discussion.

Direct measurement of the diameters of steps photographed on fast-moving cameras support this conclusion. From those photographic records which seem to be quite free from halation errors and for which the discharge was sufficiently near, I find values for r ranging from 0·5 to 5 m. Smaller values are not observed on any records. The rapid movement of the camera lens does not appear to have given the steps a false diameter, for the results are the same whether the lens moves at right angles, or parallel, to the steps.

These large values of r do not conflict with the observation (Schonland, 1937) that the radius of the channel of the *return* stroke is about 8 cm. The return streamer carries an average current of 30 000 A, 100 times as great as that in the first leader, and in consequence it must be powerfully constricted by its own magnetic field.

The intense light emitted by the narrow return streamer is no doubt the reason why the large diameter of the first leader has so far escaped observation. There is, however, evidence of such large diameters in at least one photograph by Larsen of a close air discharge (Abbot, 1934) and in records of beaded lightning. Toepler (quoted in Gockel, 1925) has reported beads in *Perlschnurblitz* which were 3·7 m in diameter.

Writers on the mechanism of the step process who have regarded the pilot streamer as a single electron avalanche have supposed its tip radius to be a centimetre or less (Schonland, 1938; Meek, 1939; Loeb and Meek, 1941). A radius of this order for the pilot streamer and hence for the leader tip behind it must be considered impossible now that it is clear that the leader tip is at a potential of between 10^7 and 10^8 V. The fields in front of the tip would be so large (10^7 to 10^8 V/cm) as to make its periodic arrest inconceivable.

5. Current density in pilot streamers

With I = 320 A and r = 2·5 m, the current density in the lightning pilot streamer is $1·6 \times 10^{-3}$ A/cm^2. The corresponding quantity in the long spark can be obtained from the results of Hagenguth *et al.* (1951). In their observations the current was recorded in the grounded lead to the anode of the gap. As the cathode pilot streamer approached this electrode the current was observed to rise to a peak of 1·5 A in one instance and 5·6 A in another.

This method of observation does not give the full current in the approaching pilot streamer except at, or just before, the moment at which it makes contact with the anode. It will therefore be assumed that the peak current observed corresponds to the continuous current I in the moving pilot streamer, though as Hagenguth has pointed out there is no way of distinguishing between the peak due to the pilot and that due to the immediately following leader streamer from the anode.

Taking I as the mean of the two observations quoted above, 3·55 A, and the observed radius of the pilot streamer, 30 cm, the mean current density becomes $1·14 \times 10^{-3}$ A/cm^2. This is in satisfactory agreement with the value $1·6 \times 10^{-3}$ A/cm^2 found above for the lightning pilot streamer.

6. Comparison of parameters of the pilot streamer in lightning and the long spark

The values of three parameters of the lightning pilot streamer, its velocity, its current density and the ratio of its length to its radius are compared with corresponding quantities for the pilot streamer of the long spark in the first three horizontal columns of *Table 1*.

Table 1

1 Parameter	2 Lightning	3 Spark	4 Reference	5 Remarks
velocity (cm/s)	$1 \cdot 5 \times 10^7$ to	$3 \cdot 0 \times 10^7$	Hagenguth *et al.* (1951)	calculated by present writer
	$8 \cdot 0 \times 10^7$	$1 \cdot 0 \times 10^7$	Gaunt and Craggs (1951)	positive streamer only
current density (A/cm²)	$1 \cdot 6 \times 10^{-3}$	$1 \cdot 14 \times 10^{-3}$	Hagenguth *et al.* (1951)	calculated by present writer
gap or step length/radius	8	10	Allibone (1948)	
		15	Hagenguth *et al.* (1951)	
		6	Gaunt and Craggs (1951)	positive streamer in nitrogen
average field strength across gap (kV/cm)		6	Hagenguth *et al.* (1951)	
		6	Gaunt and Craggs (1951)	positive streamer in nitrogen
		6 to 12	Komelkov (1947)	

The agreement in columns 2 and 3 is close enough for it to be concluded that the streamer is of the same nature and produced by the same mechanism in both lightning and the long spark.

The long spark, as shown in the last horizontal group of *Table 1* involves a further parameter, the average field strength ($\bar{X}$) which exists across the gap before breakdown takes place. When $\bar{X}$ is less than 6 kV/cm observation shows that breakdown fails to occur, because, although a streamer sets out from the electrode, it is unable to cross the gap. There are thus two field strength conditions for the advance of a spark pilot streamer; the first is that the field strength in front of its tip must exceed 30 kV/cm, the second is that the mean field inside its channel must exceed 6 kV/cm.

The reason for the second condition is to be sought in electron-capture processes taking place in the channel, which are further discussed in section 10. If a streamer is to advance, its channel must carry current mainly through the agency of free electrons. The existence of a lower limit for $\bar{X}$ indicates that in an electric field of strength below this limit, free electrons are rapidly removed by capture, and as a result most of the channel current ceases and the streamer stops.

Applied to the lightning pilot streamer this means that when at any point in its channel the field strength falls below a value of the order of 6 kV/cm, the progress of the pilot streamer is arrested. This leads (section 9) to the development of positive space charge in front of the leader and the production of a fast step streamer which overtakes the pilot and converts it into a good conductor. It will be shown in the next section that the step length calculated from this argument is in satisfactory agreement with observation.

7. The length of a step in the lightning leader process
The step length may be calculated in the following way. The pilot streamer advances by virtue of ionizing processes at its tip, made possible by a field of some

3×10^4 V/cm (section 4(*c*)). In so doing it establishes in front of the tip of the arrested leader a cylindrical region of negative charge of radius *r* and charge *q* per unit length. After the pilot has moved forward over a certain distance, this pilot charge reduces the field at the starting point of its own channel to 6 kV/cm (20 e.s.u.). At this stage the supply of current needed for its further advance is considered to cease. After a short interval needed to establish positive electrode space charge, the leader rapidly overtakes the pilot and once more comes to rest (section 9).

From the evidence discussed in section 2 it follows that to a first approximation leader and pilot carry the same charge *q* on each unit length. For this reason, although leader and pilot have very different conductivities, electrical image effects at their junction at *A* in *Figure 1* can be ignored. In *Figure 1*, *ABCD* is the temporarily arrested leader and *AE* the pilot streamer. The pilot has travelled a distance *l* by virtue of the field in front of *E*, which is of the order of 100 e.s.u. Both leader and pilot will be supposed to have the same radius *r*, an assumption which will be examined later. At any point *P* between *A* and *E* the forwardly directed (negative) field in the channel of the pilot is the resultant of a forward field due to the charge to the left of *P* and a backward field due to that to the right of *P*. The channel field is thus smallest at *A*. Applying the conclusions of section 6, the condition for the cessation of current flow into the pilot is

$$X_A \leqslant 20 \text{ e.s.u.} \qquad (2)$$

The simplest way to express X_A in terms of *l* is to assume that the cross-sectional distribution of the charge *q* per unit length is the same for both leader and pilot.

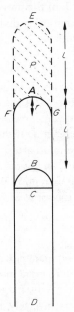

Figure 1.

Since $l \geqslant r$ any other distribution will give approximately the same result. With this assumption, it is clear from *Figure 1* that the value of X_A must be the same as that of the axial field which would be produced by the leader alone at a distance l in front of its tip, that is, if the pilot were absent and the leader tip moved back from A to B.

The axial field at a distance l in front of B is made up of two forward components, $X_A{}'$ from the charge on the hemisphere BC and $X_A{}''$ from the charge on the leader channel extending back from C to a great distance, so great that it may be taken as infinity. When $l \geqslant r$ the values of both these components become independent of the manner in which the charge q is distributed over the cross-section of the channel, and it is easily shown that $X_A{}'$ is then equal to qr/l^2 and $X_A{}''$ to $q/l + r$.

Equation (2) can therefore be written approximately

$$qr/l^2 + q/l + r \leqslant 20. \tag{3}$$

If $q = 2 \cdot 4 \times 10^4$ e.s.u./cm and $r = 2 \cdot 5$ m, this equation is satisfied if $l \geqslant 12 \cdot 5$ m. Thus current in the pilot channel at A will cease to flow when AE has extended $12 \cdot 5$ m.

It will be shown in section 9 that although current supply at A now ceases, the pilot tip E moves forward a further $1 \cdot 2$ m before it is overtaken by the bright step streamer from A. Since this step streamer will travel from the surface FAG of the leader tip, the minimum total length illuminated in the step will be

$$12 \cdot 5 + 2 \cdot 5 + 1 \cdot 2 = 16 \cdot 2 \text{ m}.$$

This result is in reasonable agreement with the value of 20 m which is observed.

Fast camera photographs have shown that each new step streamer appears to re-illuminate about 10% of the length of the previous step (Schonland *et al.*, 1935). This observation finds a ready explanation in that the new step starts from FGA and must therefore appear to re-illuminate a fraction r/l or 13% of the previous step.

It has now to be shown that the radius r of the pilot streamer is not likely to be appreciably different from that of the leader channel behind it. From section 4(*a*), $I = qv = \frac{1}{2}Xvr$, and hence $r = 2q/X$. From this it follows that unless the charge per unit length of the streamer increases, which means an increase in current, an increase in r can only be achieved by a decrease in X. If this decrease is progressive it will lead to a momentary stoppage of the pilot, after which it will proceed with its original radius. An increase in current, on the other hand, will involve an increased fall of potential along the pilot channel, reduce X, and again stop the pilot. Thus r may be expected to be stabilized at the value with which the streamer begins its development.

8. The effect of pressure upon the step parameters

The critical value of the channel field X_A used in equations (2) and (3) may be expected, like similar critical fields in gas discharges, to vary directly as the pressure. Thus since, from section 4, r must vary inversely as the pressure to satisfy the corona condition, equation (3) shows that l will also vary inversely as the pressure.

This effect should show itself in the early stages of the leader, when it is just detectable below the base of the cloud, as steps of greater length and larger radius than when the leader is near the ground. It is in fact found that the largest values of l and r are observed on the photographic records of stepped air discharges high above the ground.

These large values of l and r are, however, not likely to be due to reduced pressures only. There is evidence of considerable positive space charge in the air below the base of a thundercloud (Malan, 1953) and the type β leader with its pronounced branching and long bright steps shows clearly that local enhancements of the field guiding the pilot streamer are of considerable importance. The expression for X_A used on the left-hand side of equation (3) does not take account of this effect nor does the value of X_A on the right-hand side (20 e.s.u.) allow for a pre-existing field due to positive space charge. Since all leader movement takes place towards such space charge or towards the electrical image of the leader channel in the earth, the 'free-space' calculation of l which has been used in equation (3) must be expected to give a value somewhat lower than that usually observed and much lower than that found when the leader approaches regions of strong positive space charge.

One of the most striking features of the stepped leader process is the small range of variation of the pause time τ, which is in effect the duration of the pilot streamer stage. This small range of variation of τ is found over the whole length of individual leaders of type α, which are considered, on the evidence that they branch but little, to move in air free from strong concentrations of positive space charge. The same small range (30 to 70 μs) is exhibited by the average values of τ for different type α leaders. Over this range of τ, the value of l may vary from 10 to 80 m (Schonland *et al.*, 1935, *Figure 15*), its larger values occurring in regions of reduced pressure just below the cloud. For still larger values of l, up to 200 m, τ remains at about 80 μs.

It is not easy to interpret this small variation of τ with l. It can be shown that if the pilot tip advances by collisional ionization only so that its velocity v is the same as that of electron drift, v', τ which is equal to l/v should vary directly as l. For v' varies as $\sqrt{(X\lambda)}$, where λ is the electron mean free path and X the field. Since $X \approx 2q/r$ and $r \propto 1/p$, where p is the pressure, X should vary as p for a given value of q. Since $\lambda \propto 1/p$, v' should be independent of pressure, and if $v = v'$, $\tau \propto l$.

This conclusion, as has been shown, is not borne out by observation. It may be that v is not the same as v' because photo-ionization plays a considerable part in the advance of the pilot. Or it may be that high values of l are due always to concentrations of positive space charge in front of the tip (type-β leaders) and consequent convergence of the tip field X.

9. The bright step

The bright step stage of the leader is observed to take less than a microsecond over its journey. It travels at more than $5 \cdot 0 \times 10^9$ cm/s, the ionization in the pilot channel in front of it enabling it to reach this high velocity. Its starting-point is clearly likely to be the tip *FAG* of the arrested leader (*Figure 1*), for it is only by extending this good conductor that the current of about 1000 A (section 2) needed for its advance can be supplied. To move at high speed the step streamer requires a field of at least 3×10^4 V/cm in front of its tip. Hence X_A must rise in a few microseconds from 6×10^3 to at least 3×10^4 V/cm. The sequence of events which may lead to this change in field can be summarized as follows:

(*a*) As has been described in section 7, the field in the pilot channel immediately in front of *A* (*Figure 1*) falls to such a low value ($\leqslant 20$ e.s.u.) that free electrons are no longer available to carry current from the leader into the pilot stem. Thus the current in the region of *A* becomes vanishingly small.

(*b*) Free electrons still exist in the pilot channel further forward than *A* and these continue their forward drift by virtue of the self-inductive property of the channel. This creates positive space charge in front of the arrested leader tip *FAG*.

(*c*) The positive 'electrode' space charge builds up in this way until it has created a field in front of *FAG* which is so large that the leader itself can advance as a fast-moving step streamer. This will happen when $X_A \approx 3 \times 10^4$ V/cm $\approx 2q/r$. The positive electrode space charge has then effectively neutralized the backwardly directed field from the negatively charged stem *AE* of the pilot. The total positive electrode space charge created in this way is $Q = \pi r^2 X_A / 4\pi = \frac{1}{2}qr$. Since the pilot continues to move forward for a time t with velocity v and draws a current qv by the drift of electrons away from *A*, $Q = \frac{1}{2}qr = qvt$. Thus $t = r/2v$, or $3 \cdot 1$ μs if $r = 2 \cdot 5$ m and $v = 4 \times 10^7$ cm/s and the pilot will advance a further distance $vt = 1 \cdot 24$ m before it is overtaken by the fast step streamer. The electrode space charge in front of *A* does not itself extend over a distance as great as $1 \cdot 24$ m. The density N of free electrons in the pilot channel is $1 \cdot 8 \times 10^9 / \text{cm}^3$ (section 10), and the total charge carried on the free electrons present in 1 cm of length is $N\pi r^2 e$, or $1 \cdot 8 \times 10^5$ e.s.u. If the disappearance of free electrons from the channel in front of *A* takes place over a very small length only of *AE*, the further length required for the development of the positive charge Q ($3 \cdot 6 \times 10^6$ e.s.u.) is 20 cm.

10. Disappearance of free electrons in the pilot channel

The pilot streamer in both lightning and in the long spark has been shown in section 5 to have a current density i of about $1 \cdot 5 \times 10^{-3}$ A/cm^2. Most of this current must be carried by free electrons drifting in a channel field whose value when the pilot is fully extended has been shown to be of the order of 6×10^3 V/cm. In this field their velocity u is 5×10^6 cm/s, and the electron density $N = i/eu$ is $1 \cdot 8 \times 10^9 / \text{cm}^3$.

Two main capture processes operate upon such electrons, recombination with positive ions and the formation by attachment of O_2^- ions. The former is important only if $N\alpha t \geqslant 1$, where $\alpha = 2 \times 10^{-6}$ and t is the time during which the electrons are recombining with positive ions (Meek, 1939). In the pilot-streamer channel of lightning, $t = 50$ μs, hence $N\alpha t = 0 \cdot 18$ and loss by recombination is insignificant. This must also be true for the spark pilot, in which t is still smaller and N the same.

That loss by attachment to O_2 molecules is an important factor is clearly shown by the observation of Gaunt and Craggs (1951) that the pilot streamer from a

positive point in nitrogen is very much weakened by the addition of a small percentage of air. In pure oxygen they were not able to detect the streamer at all,
though faint glows near the electrodes indicated that it still existed in a much
enfeebled form.

It is unfortunately not possible at present to estimate the magnitude of this form
of capture loss, since the attachment cross-section found by the earlier electron
filter and Bailey-Townsend methods is 1000 times larger than that recently
reported by Biondi and Brown (1949), and no final data are available on its
variation with field strength.

Qualitatively, however, it is clear that the effect is an important one and must
increase with reduced fields acting on the electrons. Since it has been shown in
section 6 that the critical field inside the channel for the streamer advance to
continue is 20 e.s.u., it is concluded that this is the field strength below which the
majority of free electrons have too short a life to be effective carriers of current in
the pilot streamer process.

11. The subsequent leader in the long spark

Long-spark studies with impulse generators show that once the pilot streamer has
bridged a gap, bright but very narrow branched leaders enter the broad channel of
the pilot from both electrodes, travelling towards each other until they meet
(Allibone, 1938; Hagenguth *et al.*, 1951). In no instance, however, have the
diameters of these subsequent leaders been found, as in lightning, to be the same
as the diameter of the pilot streamer.

It is possible that the reason for this lies in the arrangements used to supply
current to the gap. A fast step streamer of the same nature as that observed in
lightning would require a current $I = \frac{1}{2}Xvr$, which, in the case of Hagenguth's
experiment, would be 5000 A if $X = 3 \times 10^4$ V/cm, $v = 5 \times 10^9$ cm/s and $r = 60$ cm.
Since his impulse generator of 3 MV was connected to the rod gap through a
resistance of 1530 Ω, it could not produce a current of this magnitude without a
voltage drop which would immediately stop the leader streamer. In many of the
experiments of Allibone and Meek at atmospheric pressure the series resistance
was still higher.

Dr. Allibone has, however, pointed out to me that this explanation would require
an increase in the width of the leader streamer as the series resistance was reduced,
an increase which is not observed.

An alternative explanation, for which there is some supporting evidence, is that
the longest impulsive spark breakdown studies have so far been carried out in a
transition region of the mechanism of breakdown. For applied potential differences
of 1 MV or less, breakdown may be initiated from a negative electrode by the
narrow avalanche leader only (Raether, 1937; Loeb and Meek, 1941), while for
p.d.'s of 3 MV or more, the pilot streamer may replace it as the easiest method of
breakdown. Higher voltages imply longer gaps and longer times during which the
streamer channel, pilot or avalanche, must maintain its conductivity. As has been
indicated in section 10, the pilot streamer, for which $N \approx 10^9$ electrons/cm^3, has

an advantage in this respect over the narrow avalanche streamer, for which
$N \approx 10^{14}/\text{cm}^3$ and in which electron depletion by capture will be serious when
the gap is long. The narrow avalanche is under a further disadvantage, which
increases with gap length, in that the corona effect from its surface is continually
robbing its tip of current (Bruce, 1944; Komelkov, 1950). Thus as p.d.'s, gap
lengths and breakdown durations increase, there will come a stage at which the
pilot streamer, provided the necessary current is available, can proceed with a lower
average gradient within its channel than that needed for the channel of the
avalanche streamer. At this stage it will be the preferred type of breakdown streamer.

In support of this suggestion, Hagenguth *et al.* (1951) have shown that a remark-
able change takes place in the appearance of the breakdown streamer when a
60-cycle p.d. exceeding 10^6 V is applied to a point-point gap. The corona from the
high-voltage rod was observed to change at this critical voltage from a bluish glow
to a 'very strong' streamer discharge which could be observed to advance at least
half-way across the gap. Corresponding to this change in the appearance of the
preliminary breakdown process, there set in what they describe as a radical change
in average breakdown gradient, from 5×10^3 V/cm at 10 ft spacing to $3 \cdot 3 \times 10^3$ V/cm
at 20·8 ft.

From this discussion it would seem reasonable to conclude that the mechanism
of long-spark breakdown between 10^6 and 3×10^6 V lies in a transition zone
between the electron avalanche and the pilot streamer processes, and that this is
the chief reason why when the pilot is observed the subsequent leader does not
have a large diameter.

12. Stepped-leader effects in the spark discharge
The interesting stepped leader effects observed in the spark discharge by Allibone
and Meek (1938) when the series resistance was raised to about 10^6 Ω do not seem
to correspond in any way with the steps observed in lightning leaders. They have a
low velocity (8×10^6 to $1 \cdot 5 \times 10^7$ cm/s), and there is no evidence that they are
connected with the normal pilot streamer process. It seems probable that they
occur only because the current requirements of the spark leader cannot be met by
the generator without a fall in potential so great as to stop the leader (Allibone,
1948).

13. The stepped-leader theory of Komelkov
Komelkov (1950) has put forward an explanation of the step process in lightning
which resembles that proposed here, in that the outer corona zone of the leader
channel is considered to have a large radius (≈ 10 m) but differs from it in denying
that the pilot streamer extends for any appreciable distance in front of the leader.
The stoppage of the leader is attributed by Komelkov to a fall in its tip potential
caused by the need to charge up a new section of the channel. This fall, which is
considered to amount to 5% of its original potential, has to be made good by the
flow of current from the cloud during the 50 μs pause. It has been shown in section
2 that large current pulsations in excess of 16 000 A would be required by this
theory, and that these are not observed.

Note added 6 July 1953, Liao and Anderson (1953) have shown that impulsive spark breakdown in oil, between a negative point and an earthed plane, involves a step leader process of the same nature as that found in the lightning discharge. They have been able to photograph a continuously moving pilot streamer in the intervals of each of a dozen steps.

For the oil used, the value of X in equation (1) of section 4 is 500 kV/cm and that of X_A in equation (3) of section 7 is about 32 kV/cm. By combining equations (1) and (3) it is easily shown that $r^2/l^2 + r/l = 2X_A/X$, from which $l/r = 9$. Since, as indicated in section 7, the ratio of the observed step length to the radius of the streamer is about 25% greater than l/r, this second ratio should be 11·3. The observed step length in oil is reported to vary from 0·25 to 1·25 cm, and the value of r for the leader to be about 0·045 cm, giving values for the ratio lying between 5·5 and 28, with a mean of 17·3.

Considering the uncertainties in some of these quantities, the agreement between calculated and observed values is satisfactory. It indicates that the simple theory derived to explain the step process covers values of from 20 m to a few millimetres.

My thanks are due to Dr D. J. Malan for helpful discussions and to him and Mr N. D. Clarence as well as to Professor D. B. Hodges for permission to use some of their unpublished results. I wish also to thank Professor L. B. Loeb of the University of California, with whom I discussed an early draft of the ideas in this paper, for his encouragement, and Dr T. E. Allibone, F.R.S., for helpful advice on long-spark problems.

References

Abbot, C. G., *Smithson. Misc. Coll.* **92**, no. 12, (1934).

Allibone, T. E., *Nature* **161**, 970, (1948).

Allibone, T. E. and Meek, J. M., *Proc. Roy. Soc.* A, **166**, 97; **169**, 246, (1938).

Biondi, M. and Brown, S. G., *Phys. Rev.* **76**, 1697, (1949).

Bruce, C. E., *Proc. Roy. Soc.* A, **183**, 228, (1944).

Chapman, F. W., *Proc. Phys. Soc.* **51**, 876, (1939).

Cravath, A. and Loeb, L. B., *Physics* **6**, 125, (1935).

Gaunt, H. M. and Craggs, J. D., *Nature* **167**, 860, (1951).

Gockel, A., *Das Gewitter* 63. F. Dümmlers, Verlag, (1925).

Goodlet, B. L., *J. Instn Elect. Engrs.* **81**, 1, (1937).

Hagenguth, J. H. and Anderson, J. G., *Trans. Amer. Inst. Elect. Engrs.* **71**, 641, (1952).

Hagenguth, J. H., Rohlfs, A. F. and Degnan, W. J., *Trans. Amer. Inst. Elect. Engrs.* **71**, 455, (1951).

Komelkov, V. S., *Bull. Acad. Sci. U.S.S.R.* No. 8, 955; *C.R. Acad. Sci.* U.S.S.R. **68**, 57, (1947).

Komelkov, V. S., *Bull. Acad. Sci.* U.S.S.R. No. 6, 851, (1950).

Liao, T. W. and Anderson, J. G., *A.I.E.E. Tech. pap.* 53-280, (1953).

Loeb, L. B. and Meek, J. M., *The mechanism of the electric spark.* Stanford University Press, (1941).

Malan, D. J., *Ann. Géophys.* 8 (4), 1, (1953).

Malan, D. J.,and Schonland, B. F. J., *Proc. Roy. Soc.* A, 191, 489, (1947).

Malan, D. J. and Schonland, B. F. J., *Proc. Roy. Soc.* A, 209, 158, (1951).

McEachron, K. B., *J. Franklin Inst.* 227, 149, (1939).

Meek, J. M., *Phys. Rev.* 55, 972, (1939).

Raether, H., *Z. Phys.* 107, 93, (1937).

Rudenberg, R., *Wiss. Veroff. Siemens-Konz.* 9, pt. 1, (1930).

Schonland, B. F. J., *Phil. Mag.* 23, 503, (1937).

Schonland, B. F. J., *Proc. Roy. Soc.* A, 164, 132, (1938).

Schonland, B. F. J., Hodges, D. B. and Collens, H., *Proc. Roy. Soc.* A, 166, 56, (1938).

Schonland, B. F. J., Malan, D. J. and Collens, H., *Proc. Roy. Soc.* A, 152, 595, (1935).

Szpor, S., *Bull. Ass. Suisse El.* 1, 1, (1942).

Wilson, C. T. R., *Phil. Trans.* A, 221, 73, (1920).

Introduction to Papers 24 and 25 (Meyerand and Haught, 'Gas breakdown at optical frequencies'; Minck, 'Optical frequency electrical discharges in gases')

The first report of gaseous breakdown by focussed high intensity laser radiation was that by Maker and his colleagues at the Third International Conference on Quantum Electronics in 1963. Between then and the time of the Fourth Conference (San Juan, Puerto Rico, 1965) at least fifteen other papers appeared in which the breakdown processes involved were discussed, and there were seven additional papers at the Fourth Conference itself. Interest in the subject has continued to increase in the past few years and a number of surveys and bibliographies have already been published (e.g. that by De Michelis, referenced below).

It is appropriate in a collection of papers concerned with electrical breakdown to limit ourselves to considering the processes which occur between the laser flash and the attainment of an electron concentration of say 10^{15} electrons per cubic centimetre, i.e. the initiation and growth stages of the discharge. There appear to have been for some years considerable differences of opinion concerning the breakdown mechanisms involved but the general consensus of opinion now seems to favour, as the initiatory process, multiphoton photoionisation of the gas (or gasous impurities) followed by an avalanche growth of ionisation. The avalanche process is taken as involving the absorption of photons by electrons in free-free transitions (also called 'inverse bremsstralung absorption') and subsequent collisional excitation and ionisation and perhaps, photo-ionisation (see the paper by A. V. Phelps mentioned below). The confusion in the literature over the merits of avalanche mechanisms based respectively on free-free absorption or on an extension of the theory of microwave breakdown to optical frequencies appears to have been resolved in 1965 when it was pointed out by Phelps that for photon energies, $h\nu$, much less than the energy of the electrons, the Boltzmann equation describing the absorption of finite photon energies reduces to the microwave limit of the Boltzmann equation used to describe the process of microwave electrical breakdown. Similar points were made by P. F. Browne, and A. F. Haught, R. G. Meyerand, and D. C. Smith.

The spatial and temporal growth of ionisation in the avalanche process are, of course, affected by loss mechanisms such as diffusion and recombination, and also by electron attachment in certain gases. A recent paper in which solutions of the

continuity equation appropriate to various experimental conditions are considered
is that by Morgan *et al.*

The two short papers reproduced here were among the very earliest on laser-
induced electrical breakdown and describe some of the main features observed.
They both invoke avalanche mechanisms to explain their observations and introduce
ideas expanded by later workers. Other early papers which deserved inclusion and
should be consulted are those by J. K. Wright, A. Gold and H. B. Bebb, B. A. Tozer,
and L. V. Keldysh.

Further Reading

Maker, P. D., Terhune, R. W. and Savage, C. M. in "Quantum Electronics III"
     (Dunod, Paris, 1964).

Tozer, B. A., *Phys. Rev.* **137**, A1665, (1965).

Wright, J. K., *Proc. Phys. Soc.* **84**, 41, (1964).

Gold, A. and Bebb, H. B., *Phys. Rev. Letters* **14**, 60, (1965).

Browne, P. F., *Proc. Phys. Soc.* **86**, 1323, (1965).

Keldysh, L. V., *Soviet Physics J.E.T.P.* **20**, 1307, (1965).

Haught, A. F., Meyerand, R. G. and Smith, D. C., in *Physics of Quantum Electronics*,
     McGraw-Hill, 1966.

Phelps, A. V., in *Physics of Quantum Electronics*, McGraw-Hill, 1966.

→De Michelis, C., *IEEE J. Quantum Electron.* **QE-5**, 188, (1969).

Morgan, F., Evans, L. R. and Grey Morgan, C., *J. Phys. D.* **4**, 225, (1971).

24

Gas breakdown at optical frequencies
R. G. MEYERAND, JR., AND A. F. HAUGHT*

A Q-spoiled ruby-laser system capable of generating a giant pulse of optical energy
with a peak power of the order of tens of megawatts has been used to study the
interaction of extremely high-intensity optical-frequency electromagnetic radiation
with gases. The laser, shown in *Figure 1*, consists of a $\frac{1}{2}$-in. diameter, 6-in. long
ruby rod pumped by four E. G. & G. FX-47 lamps, each lamp individually
powered by a 1200-μF capacitor bank. A polarizer–Kerr cell shutter was used
to alter the Q of the laser cavity resulting in a single giant pulse of optical
radiation from the laser.[1] The light emitted by the laser is incident on a lens which
forms one window of a cell containing the test gas. At the focus of the lens,
breakdown of the test gas by the focused laser beam is observed for suitable
conditions of beam energy and gas pressure.[2,3]

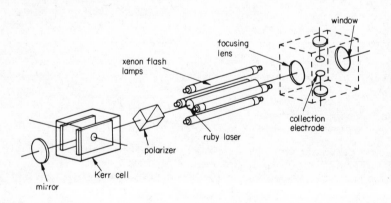

Figure 1. Q-spoiled laser system.

* Reproduced from *Physical Review Letters,* **11**, 401, (1963) by permission of The American
Physical Society, New York.

The breakdown itself is evidenced by the appearance of a bright flash of light at the focus of the lens. The light is not simply scattered laser radiation, since it is readily seen through 'band-block' interference filters designed to exclude the 6934 Å ruby-laser output, and in addition, from photomultiplier and high-speed framing camera photographs, the breakdown luminosity lasts for a time of the order of 50 μs compared with the 30-nsec duration of the exciting giant laser pulse.

A pair of electrodes placed on either side of the focus point were used to establish that electrical breakdown, that is, the production of ion pairs, was indeed achieved. Neither the current wave form nor the total charge collected, about 10^{13} electron charges, was affected by electrode potential differences from 100 V to 200 V, the range of the power supply used. Photomultiplier records show that the laser pulse and the breakdown occur simultaneously within less than 50 nsec, the time resolution of the dual-beam oscilloscope.

To compare the phenomena observed here with existing theories of electrical breakdown in gases, measurements have been made of the optical-frequency electrical field required for breakdown as a function of pressure for a number of gases. The electric field strength at the point of breakdown was determined from calorimetric measurements of the energy in the giant optical pulse, the time duration of the pulse as determined by photomultiplier records, and the focus diameter. For a typical case, the energy of the giant pulse was one joule and its duration 30 nsec, giving a peak power of 30 megawatts. The diameter of the focus point, determined by a series of measurements of the hole size produced in 5×10^{-6}-cm thick gold foil, was 2×10^{-2} cm resulting in a peak power density of about 10^{15} W/m^2. This focus diameter was confirmed by calculations of the demagnification ratio of the optics using the measured laser-beam divergence. Both of these techniques give an optical-frequency electric field strength of approximately 10^7 V/cm at the focus.

The field strength necessary to produce breakdown in argon and helium is shown in *Figure 2* as a function of gas pressure. In each case the breakdown threshold is observed to decrease with increasing pressure, levelling off at the higher pressures. The field strengths required for breakdown in argon and helium are less than 10^7 V/cm, or 0·1 V across the dimensions of an atom. It should be noted that this field strength is less by two orders of magnitude than that required for a direct electric field stripping of an electron from an atom.

Similarly, since the laser photon energy is approximately 1·7 eV and gases with an ionization potential as high as 24 V (helium) have been successfully ionized by the laser beam, direct single-step photoionization is not possible. Successive photon absorption could conceivably produce the ionization. However, to reach even the lowest lying levels of argon and helium would require, respectively, 7 and 12 successive absorptions. Assuming even a 10% probability for each absorption step, this would imply a 10^5 ratio between the field strengths required for breakdown in argon and helium. The experimentally observed ratio is only 1·7, indicating that multiple photon absorption is not responsible for the breakdown.

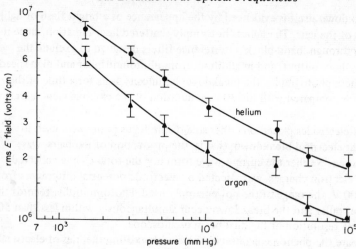

Figure 2. Breakdown field strength as a function of pressure.

The curves of *Figure 2* are threshold curves, that is, for values of pressure and field strength which lie below the curve for a given gas, no ionization is observed in that gas, while at a slightly higher pressure or field strength above the curve, breakdown results and a very large degree of ionization is produced. This abrupt change implies some form of cascade process with a critical level of energy input required to sustain the process, a condition not predicted by the photon-atom processes considered above.

Cascade theories have been developed for gas breakdown at microwave frequencies.[4] The theories predict an energy input to the electrons from the oscillating electromagnetic field through the mechanism of electron-atom collisions which convert the ordered oscillatory motion of the electrons in the field to random motion. The electrons gain random energy on each collision until they are able to make ionizing collisions with gas atoms leading to succeeding generations of electrons. These theories, however, cannot be applied to explain the breakdown at optical frequencies, since important differences exist between the two cases. In the microwave case, the maximum kinetic energy in the ordered oscillatory motion of the electron in the electromagnetic field corresponds to a kinetic energy of the order of 10^{-3} eV, while typical microwave photon energies are of the order of 10^{-5} to 10^{-6} eV. Thus, an electron oscillating in the electromagnetic field must absorb and emit many microwave quanta of energy every cycle, and its motion may be considered to be classical. In the optical case, for the field strengths used in these experiments, the oscillatory energy of the electron is classically also of the order of 10^{-3} eV, but the photon energy for a ruby laser is $1 \cdot 7$ eV. The classical oscillatory energy of the electron is thus small compared with the photon energy, and one might expect that the motion of the electron and its subsequent interaction with atoms and the radiation field is governed by quantum effects. In fact, from the uncertainty principle the energy of the electron during any one cycle of the oscillating electromagnetic field cannot be determined to better than $\Delta E \geqslant \hbar v$, and therefore, it is not meaningful to discuss an electron oscillation energy of 10^{-3} eV in a 1·7-eV photon field.

Even assuming that the electron behaves classically in an optical-frequency field, not enough energy is given to the electrons to account for the ionization observed. At a helium pressure of 1500 mm Hg and an optical field strength at which breakdown occurs, microwave breakdown theory predicts that an average electron would gain only 190 eV of energy during the optical pulse. Thus assuming 30 eV are required to produce an ion pair, the ionization produced during the pulse would be many orders of magnitude below the experimentally observed value of 10^{13} ion pairs, determined by the charge collection experiments.

In the classical theories for breakdown at microwave frequencies, an electron can receive at most only the 10^{-3} eV of ordered oscillatory energy from the electric field in an electron-atom collision. However, at optical frequencies the absorption can become a quantum-dominated process, and the possibility exists that a large fraction of the photon energy can be transferred to an electron during a collision. The process of bremsstrahlung, photon radiation from electrons during an interaction with an atom or ion, is well known. The reverse process may also occur in which an electron gains energy by absorbing either the entire photon or some significant fraction of its energy during a collision with an atom.[5] It is felt that a process of this inverse bremsstrahlung type is the mechanism responsible for the breakdown at optical frequencies.

The authors would like to express their appreciation to Professor S. C. Brown of the Massachusetts Institute of Technology, Cambridge, Massachusetts for many helpful discussions and comments during the course of this work.

References

1. McClung, F. J. and Hellwarth, R. W., *Proc. I.E.E.E.* **51**, 46, (1963).
2. Terhune, R. W., Third International Symposium on Quantum Electronics, Paris, February 1963 (to be published).
3. Damon, E. K. and Tomlinson, R. G., *Appl. Opt.* **2**, 546, (1963).
4. Brown, S. C., *Handbuch der Physik* (Springer-Verlag, Berlin, 1956), pp. 531–574.
5. Bethe, H. A. and Salpeter, E. E., *Quantum Mechanics of One- and Two-Electron Atoms* (Springer-Verlag, Berlin, 1957), pp. 317-335.

25

Optical frequency electrical discharges in gases

R. W. MINCK*

By focusing the output beam of a giant-pulse laser, electrical discharge phenomena have been observed in air.[1] The purpose of this communication is to present data on breakdown for various gases and to show that the behaviour can be predicted by an extension of microwave frequency discharge theory. As with other sparks, a brilliant flash is seen and a sharp sound is heard. The spark is presumably initiated at the focus but quickly grows to a volume several mm in length and 1 mm in diameter. The results of this work indicate that the maximum cw power density that can be transmitted through the air is 7×10^{11} W/cm^2 at all pressures for which the electron mean free path is small compared to the beam diameter.

Triangular shaped pulses with 25-nsec half-widths and peak powers of 3 to 5 MW were produced by a Q-spoiled ruby laser employing a rotating prism. The parallel beam after passing through the flat window of a pressure cell was focused by a 2-cm focal length lens. The stainless steel cell was designed for 100 atm with glass entrance and exit windows and a side port with a quartz window through which the discharge was viewed. Spectra were obtained by focusing the light from the side port on the entrance slit of a 2-m Littrow spectrograph and recording on Kodak 103F plates.

The pressure dependence of threshold for several gases is summarized in *Figure 1*, which demonstrates that the threshold decreases with increasing pressures for all the gases. Threshold is defined as that power level at which a visible spark was obtained. The actual threshold appeared to be very sharp as no weak sparks were observed. The uncertainty in the data is probably due to the variation of the fields at the focus associated with different spatial characteristics of each laser pulse.

The pressure dependences of the rare gases are notably different from those for the diatomic gases. The much steeper slopes lead to exceptionally low power levels for breakdown at high pressures. The sharp break in slope exhibited by helium near 30 atm leads to a threshold nearly independent of pressure.

* Reproduced from *Journal of Applied Physics*, **35**, 252, (1964) by permission of The Institute of Physics and The Physical Society, London.

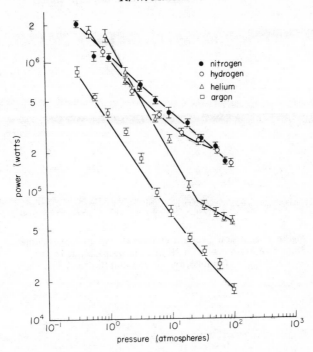

Figure 1. Pressure dependence of threshold power levels for discharge in various gases.

Emission spectra are shown in *Figure 2* for nitrogen and helium. Lines characteristic of singly ionized molecules and atoms are seen, but no lines corresponding to a second ionization have yet been observed. The line broadening could be due to electron-atom collisions. When the initial pressure was increased to 15 atm or higher, only a continuum was observed. The time dependence of the emission from the spark in air was observed with a photomultiplier. An initial peak with a shape closely resembling the laser pulse shape contained about half the total spark energy. It was followed by an afterglow decaying in about 150 nsec.

The theory for discharge at microwave frequencies[2, 3] has been extrapolated to the optical frequency range and applied to nitrogen. Ionization is produced by those electrons which have acquired a kinetic energy exceeding the ionization potential. Because of the high frequency, the peak kinetic energy achieved by an electron during a cycle is only a small fraction of an eV. Higher energies can be achieved if a collision reduces the electron velocity in the direction of the field about the time when the field reverses direction. The problem has been analyzed in terms of random walk probabilities to yield a distribution of electron energies. To achieve breakdown, the ionization rate must exceed electron losses from diffusion and attachment by an amount sufficient to allow buildup during the duration of the pulse.

Figure 2. Emission spectra from laser-excited discharges in the 5000- to 4000-A region for N_2 at 1 atm (top), N_2 at 15 atm (middle), and He at 1 atm (bottom).

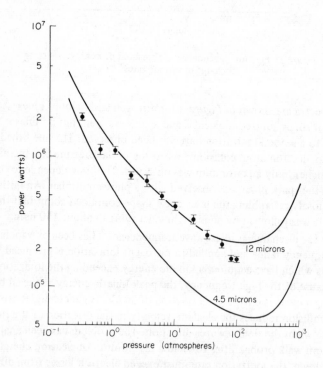

Figure 3. Comparison of calculated threshold in air for theoretical focal diameter and a probable focal diameter with observed data for N_2.

In *Figure 3* the experimental data are compared with the predicted behaviour for both an ideal focus and a focus degraded by lens aberration and laser characteristics. If attachment were the only electron loss mechanism, the short pulse duration would slightly modify the threshold. However, for the dimensions of an optical focus, the dominant loss is by diffusion, making the quality of focus very important in determining the threshold level. The calculated curves in *Figure 3* are limited by diffusion at low pressures and by collision frequency at high pressures. The break in the slope for helium is thought to be due to the short pulse time.

Further investigation is required in the high pressure region because to achieve electron collision frequency greater than optical frequency, the density of the gas must be nearly that of a liquid. The approximations in the theory for gas breakdown become very suspect under these conditions. It has been observed that liquids and solids break down at power levels comparable to those required for gases near 100 atm.

At optical frequencies another ionization mechanism exists. The probability is quite high for double and triple quantum absorption from the first excited level of the molecule to higher excited levels making it unnecessary for the electron to achieve ionization energy but only enough energy to produce the first excitation. Once the molecule is excited, the intense optical frequency electrical fields can then complete the ionization. The close agreement of the microwave theory is taken as indication that such multiple photon absorption processes are not important for nitrogen.

References

1. Maker, P. D., Terhune, R. W. and Savage, C. M., 'Optical Third Harmonic Generation in Various Solids and Liquids,' Proceedings of the Third International Quantum Electronics Conference, 1963.
2. Scharfman, W. E., Taylor, W. C. and Morita, T., 'Research Study of Microwave Breakdown of Air at High Attitudes,' Stanford Research Institute AFCRL-62-732.
3. Gould, L. and Roberts, L. W., *J. Appl. Phys.* 27, 1162, (1956).